Azure Data Engineering Cookbook

Design and implement batch and streaming analytics
using Azure Cloud Services

Ahmad Osama

BIRMINGHAM—MUMBAI

Azure Data Engineering Cookbook

Group Product Manager: Kunal Parikh
Publishing Product Manager: Reshma Raman
Senior Editor: Roshan Ravikumar
Content Development Editor: Athikho Sapuni Rishana
Technical Editor: Manikandan Kurup
Copy Editor: Safis Editing
Project Coordinator: Aishwarya Mohan
Proofreader: Safis Editing
Indexer: Priyanka Dhadke
Production Designer: Roshan Kawale

First published: April 2021
Production reference: 1100321

Published by Packt Publishing Ltd.
Livery Place
35 Livery Street
Birmingham
B3 2PB, UK.

ISBN 978-1-80020-655-7

www.packt.com

Contributors

About the author

Ahmad Osama works for Pitney Bowes Pvt Ltd. as a database engineer and is a Microsoft Data Platform MVP. In his day-to-day job at Pitney Bowes, he works on developing and maintaining high performance on-premises and cloud SQL Server OLTP environments, building CI/CD environments for databases and automation. Outside his day job, he regularly speaks at user group events and webinars conducted by the DataPlatformLabs community.

About the reviewers

Sawyer Nyquist is a consultant based in Grand Rapids, Michigan, USA. His work focuses on business intelligence, data analytics engineering, and data platform architecture. He holds the following certifications from Microsoft: MCSA BI Reporting, Data Analyst Associate, and Azure Data Engineer Associate. Over his career, he has worked with dozens of companies to strategize and implement data analytics, and technology to drive growth. He is passionate about delivering enterprise data analytics solutions by building ETL pipelines, designing SQL data warehouses, and deploying modern cloud technologies for custom dashboards and reporting.

Has Altaiar is a software engineer at heart and a consultant by trade. He lives in Melbourne, Australia, and is the Executive Director at vNEXT Solutions. His work focuses on data, IoT, and AI on Microsoft Azure, and two of his latest IoT projects won multiple awards. Has is also a Microsoft Azure MVP and a regular organizer and speaker at local and international conferences, including Microsoft Ignite, NDC, and ServerlessDays. You can follow him on Twitter at @hasaltaiar

Table of Contents

Preface

1

Working with Azure Blob Storage

2

Working with Relational Databases in Azure

3

Analyzing Data with
Azure Synapse Analytics

4

Control Flow Activities in Azure Data Factory

5

Control Flow Transformation and the Copy Data Activity in Azure Data Factory

6

Data Flows in Azure Data Factory

7
Azure Data Factory Integration Runtime

8
Deploying Azure Data Factory Pipelines

9
Batch and Streaming Data Processing with Azure Databricks

Other Books You May Enjoy

Index

Preface

Data engineering is a growing field that focuses on preparing data for analysis. This book uses various Azure services to implement and maintain infrastructure to extract data from multiple sources and then transform and load it for data analysis.

This book takes you through different techniques for performing big data engineering using Microsoft cloud services. It begins by showing you how Azure Blob storage can be used for storing large amounts of unstructured data and how to use it for orchestrating a data workflow. You'll then work with different Cosmos DB APIs and Azure SQL Database. Moving on, you'll discover how to provision an Azure Synapse database and find out how to ingest and analyze data in Azure Synapse. As you advance, you'll cover the design and implementation of batch processing solutions using Azure Data Factory, and understand how to manage, maintain, and secure Azure Data Factory pipelines. You'll also design and implement batch processing solutions using Azure Databricks and then manage and secure Azure Databricks clusters and jobs. In the concluding chapters, you'll learn how to process streaming data using Azure Stream Analytics and Data Explorer.

By the end of this Azure book, you'll have gained the knowledge you need to be able to orchestrate batch and real-time **extract, load, transform (ETL)** workflows in Microsoft Azure.

Who this book is for

This book is for database administrators, database developers, and ETL developers looking to build expertise in Azure data engineering using a recipe-based approach. Technical architects and database architects with experience in designing data or ETL applications either on-premises or on any other cloud vendor who want to learn Azure data engineering concepts will also find this book useful. Prior knowledge of Azure fundamentals and data engineering concepts is required.

What this book covers

Chapter 1, Working with Azure Blob Storage, covers how to work with Azure Blob storage and understand how it is used when orchestrating a data workflow.

Chapter 2, Working with Relational Databases in Azure, explains how to provision and work with Azure SQL Database.

Chapter 3, Analyzing Data with Azure Synapse Analytics, describes how to provision an Azure Synapse database and ingest and analyze data in Azure Synapse.

Chapter 4, Control Flow Activities in Azure Data Factory, explains how to implement different control activities available in Azure Data Factory.

Chapter 5, Control Flow Transformation and the Copy Data Activity in Azure Data Factory, explains how to work with the Azure Data Factory integration runtime. You'll also learn to use the SSIS package with Azure Data Factory.

Chapter 6, Data Flow in Azure Data Factory, explains how to use Azure Data Factory mapping and wrangling data flow to extract, transform, and load data.

Chapter 7, Azure Data Factory Integration Runtime, details the different integration runtimes available and their use cases.

Chapter 8, Deploying Azure Data Factory Pipelines, describes how to manually and automatically deploy Azure Data Factory pipelines using the Azure portal and Azure DevOps, respectively.

Chapter 9, Batch and Streaming Data Processing with Azure Databricks, covers recipes to perform batch and streaming data processing using Azure Databricks.

To get the most out of this book

You'll need an Azure subscription along with prior knowledge of Azure fundamentals and data engineering concepts.

Software/hardware covered in the book	OS requirements
Azure Subscription	Windows, macOS X, and Linux (any)

If you are using the digital version of this book, we advise you to type the code yourself or access the code via the GitHub repository (link available in the next section). Doing so will help you avoid any potential errors related to the copying and pasting of code.

Download the example code files

You can download the example code files for this book from GitHub at `https://github.com/PacktPublishing/azure-data-engineering-cookbook`. In case there's an update to the code, it will be updated on the existing GitHub repository.

We also have other code bundles from our rich catalog of books and videos available at `https://github.com/PacktPublishing/`. Check them out!

Download the color images

We also provide a PDF file that has color images of the screenshots/diagrams used in this book. You can download it here: `https://static.packt-cdn.com/downloads/9781800206557_ColorImages.pdf`.

Conventions used

There are a number of text conventions used throughout this book.

`Code in text`: Indicates code words in text, database table names, folder names, filenames, file extensions, pathnames, dummy URLs, user input, and Twitter handles. Here is an example: "The `orders.txt` file doesn't have headers, so column names will be assigned as `_col1_`, `_col2_`, and so on."

A block of code is set as follows:

```
#get blob reference
$blobs = Get-AzStorageBlob -Container $destcontainername
-Context $storagecontext
#change the access tier of all the blobs in the container
$blobs.icloudblob.setstandardblobtier("Cool")
#verify the access tier
$blobs
```

Bold: Indicates a new term, an important word, or words that you see on screen. For example, words in menus or dialog boxes appear in the text like this. Here is an example: "The **Performance** tier can be either **Standard** or **Premium**."

> **Tips or important notes**
> Appear like this.

Sections

In this book, you will find several headings that appear frequently (*Getting ready*, *How to do it...*, *How it works...*, *There's more...*, and *See also*).

To give clear instructions on how to complete a recipe, use these sections as follows:

Getting ready

This section tells you what to expect in the recipe and describes how to set up any software or any preliminary settings required for the recipe.

How to do it...

This section contains the steps required to follow the recipe.

How it works...

This section usually consists of a detailed explanation of what happened in the previous section.

There's more...

This section consists of additional information about the recipe in order to make you more knowledgeable about the recipe.

See also

This section provides helpful links to other useful information for the recipe.

Get in touch

Feedback from our readers is always welcome.

General feedback: If you have questions about any aspect of this book, mention the book title in the subject of your message and email us at customercare@packtpub.com.

Errata: Although we have taken every care to ensure the accuracy of our content, mistakes do happen. If you have found a mistake in this book, we would be grateful if you would report this to us. Please visit www.packtpub.com/support/errata, selecting your book, clicking on the Errata Submission Form link, and entering the details.

Piracy: If you come across any illegal copies of our works in any form on the internet, we would be grateful if you would provide us with the location address or website name. Please contact us at copyright@packt.com with a link to the material.

If you are interested in becoming an author: If there is a topic that you have expertise in and you are interested in either writing or contributing to a book, please visit `authors.packtpub.com`.

Reviews

Please leave a review. Once you have read and used this book, why not leave a review on the site that you purchased it from? Potential readers can then see and use your unbiased opinion to make purchase decisions, we at Packt can understand what you think about our products, and our authors can see your feedback on their book. Thank you!

For more information about Packt, please visit `packt.com`.

1
Working with Azure Blob Storage

Azure Blob storage is a highly scalable and durable object-based cloud storage solution from Microsoft. Blob storage is optimized to store large amounts of unstructured data such as log files, images, video, and audio.

It is an important data source in structuring an Azure data engineering solution. Blob storage can be used as a data source and destination. As a source, it can be used to stage unstructured data, such as application logs, images, and video and audio files. As a destination, it can be used to store the result of a data pipeline.

In this chapter, we'll learn to read, write, manage, and secure Azure Blob storage and will cover the following recipes:

- Provisioning an Azure storage account using the Azure portal
- Provisioning an Azure storage account using PowerShell
- Creating containers and uploading files to Azure Blob storage using PowerShell
- Managing blobs in Azure Storage using PowerShell
- Managing an Azure blob snapshot in Azure Storage using PowerShell

- Configuring blob life cycle management for blob objects using the Azure portal
- Configuring a firewall for an Azure storage account using the Azure portal
- Configuring virtual networks for an Azure storage account using the Azure portal
- Configuring a firewall for an Azure storage account using PowerShell
- Configuring virtual networks for an Azure storage account using PowerShell
- Creating an alert to monitor an Azure storage account
- Securing an Azure storage account with SAS using PowerShell

Technical requirements

For this chapter, the following are required:

- An Azure subscription
- Azure PowerShell

The code samples can be found at `https://github.com/PacktPublishing/azure-data-engineering-cookbook`.

Provisioning an Azure storage account using the Azure portal

In this recipe, we'll provision an Azure storage account using the Azure portal. Azure Blob storage is one of the four storage services available in Azure Storage. The other storage services are **Table**, **Queue**, and **file share**.

Getting ready

Before you start, open a web browser and go to the Azure portal at `https://portal.azure.com`.

How to do it...

The steps for this recipe are as follows:

1. In the Azure portal, select **Create a resource** and choose **Storage account – blob, file, table, queue** (or, search for `storage accounts` in the search bar. Do not choose **Storage accounts (classic)**).

2. A new page, **Create storage account**, will open. There are five tabs on the
    Create storage account page – **Basics**, **Networking**, **Advanced**, **Tags**, and
    Review + create.

3. In the **Basics** tab, we need to provide the Azure **Subscription**, **Resource group**,
    Storage account name, **Location**, **Performance**, **Account kind**, **Replication**, and
    Access tier values, as shown in the following screenshot:

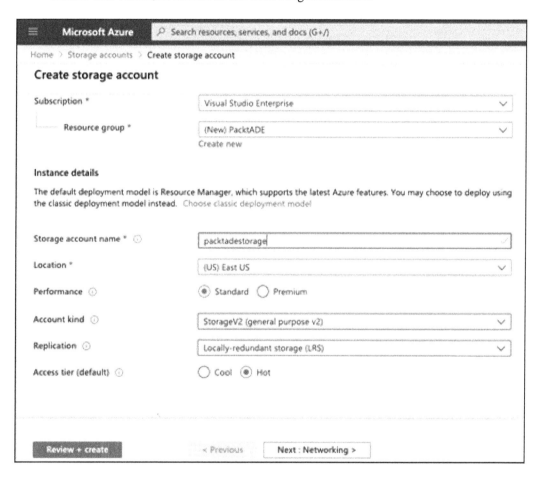

Figure 1.1 – The Create storage account Basics tab

4.   In the **Networking** tab, we need to provide the connectivity method:

Figure 1.2 – Create storage account – Networking

5. In the **Advanced** tab, we need to select the **Security**, **Azure Files**, **Data protection**, and **Data Lake Storage Gen2** settings:

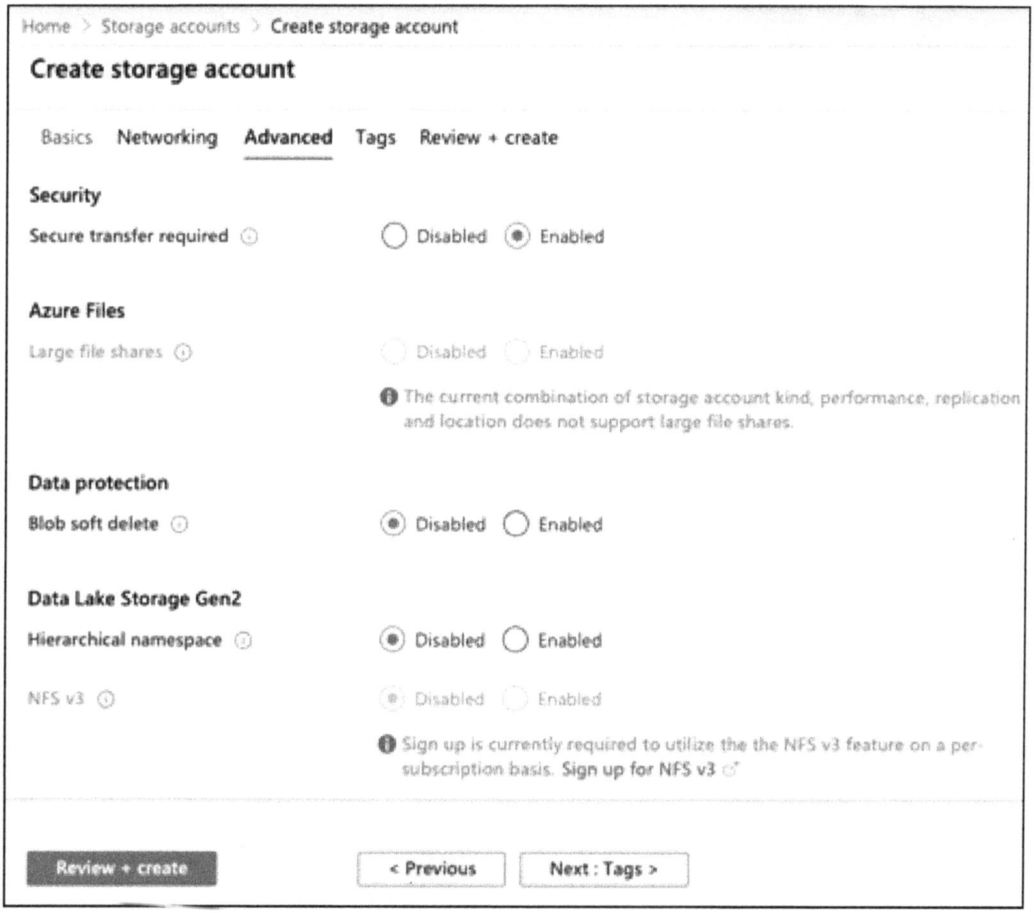

Figure 1.3 – Create storage account – Advanced

6.   In the **Review + create** tab, review the configuration settings and select **Create** to provision the Azure storage account:

Figure 1.4 – Create storage account – Review + create

How it works...

The Azure storage account is deployed in the selected subscription, resource group, and location. The **Performance** tier can be either **Standard** or **Premium**. A **Standard** performance tier is a low-cost magnetic drive-backed storage. It's suitable for applications such as static websites and bulk storing flat files. The **Premium** tier is a high-cost SSD-backed storage service. The **Premium** tier can only be used with Azure virtual machine disks for I/O-intensive applications.

Account kind is of two types, **general purpose** (StorageV1 and StorageV2) and **Blob storage**. General purpose provides one account to manage multiple Azure storage services, such as blobs, files, disks, queues, and tables. The Blob storage account only supports block blob and append blob storage. It doesn't support blobs, tables, or queues.

There are six replication types available for different high availability and durability requirements. **Locally redundant storage** (**LRS**) is the cheapest and minimal durability option. The data is replicated three times within the same data center in the primary region.

Azure storage accounts can be accessed publicly over the internet, publicly through selected networks (selected IPs, IP ranges), and from private endpoints.

Provisioning an Azure storage account using PowerShell

PowerShell is a scripting language used to programmatically manage various tasks. In this recipe, we'll learn to provision an Azure storage account using PowerShell.

Getting ready

Before you start, we need to log in to the Azure subscription from the PowerShell console. To do this, execute the following command in a new PowerShell window:

```
Connect-AzAccount
```

Then, follow the instructions to log in to the Azure account.

How to do it...

The steps for this recipe are as follows:

1. Execute the following command in a PowerShell window to create a new resource group. If you want to create the Azure storage account in an existing resource group, this step isn't required:

```
New-AzResourceGroup -Name Packtade-powershell -Location
'East US'
```

You should get the following output:

```
ResourceGroupName : Packtade-powershell
Location          : eastus
ProvisioningState : Succeeded
Tags              :
ResourceId        : /subscriptions/b85b0984-a391-4f22-a832-fb6e46c39f38/resourceGroups/Packtade-powershell
```

Figure 1.5 – Creating a new resource group

2. Execute the following command to create a new Azure storage account in the `Packtade-powershell` resource group:

```
New-AzStorageAccount -ResourceGroupName Packtade-
powershell -Name packtstoragepowershell -SkuName
Standard_LRS -Location 'East US' -Kind StorageV2
-AccessTier Hot
```

You should get the following output:

```
StorageAccountName     ResourceGroupName   PrimaryLocation SkuName      Kind      AccessTier CreationTime          ProvisioningState EnableHttpsTrafficOnly LargeFileSha
                                                                                                                                                            res
------------------     -----------------   --------------- -------      ----      ---------- ------------          ----------------- ---------------------- ------------
packtstoragepowershell Packtade-powershell eastus          Standard_LRS StorageV2 Hot        3/26/2020 8:02:57 AM Succeeded          True
```

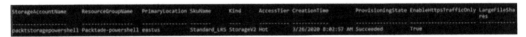

Figure 1.6 – Creating a new storage account

How it works...

There's a single command to create an Azure storage account using PowerShell – `New-AzStorageAccount`. The `SkuName` parameter specifies the performance tier and the `Kind` parameter specifies the account kind. In the later recipes, we'll look at how to assign public/private endpoints to an Azure storage account using PowerShell.

Creating containers and uploading files to Azure Blob storage using PowerShell

In this recipe, we'll create a new container and will upload files to Azure Blob storage using PowerShell.

Getting ready

Before you start, perform the following steps:

1. Make sure you have an existing Azure storage account. If not, create one by following the *Provisioning an Azure storage account using PowerShell* recipe.

2. Log in to your Azure subscription in PowerShell. To log in, run the
    `Connect-AzAccount` command in a new PowerShell window and follow
    the instructions.

How to do it...

The steps for this recipe are as follows:

1. Execute the following commands to create the container in an Azure storage account:

    ```
    $storageaccountname="packtadestorage"
    $containername="logfiles"
    $resourcegroup="packtade"
    #Get the Azure Storage account context
    $storagecontext = (Get-AzStorageAccount -ResourceGroupName
    $resourcegroup -Name $storageaccountname).Context;
    #Create a new container
    New-AzStorageContainer -Name $containername -Context
    $storagecontext
    ```

    Container creation is usually very quick. You should get the following output:

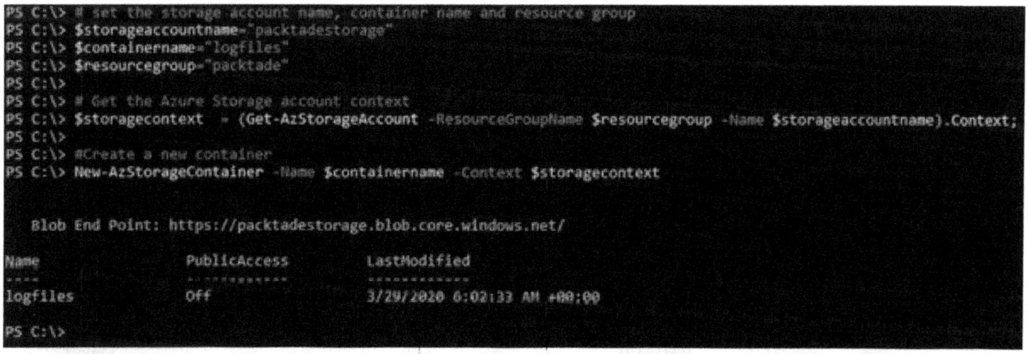

    ```
    PS C:\> # set the storage account name, container name and resource group
    PS C:\> $storageaccountname="packtadestorage"
    PS C:\> $containername="logfiles"
    PS C:\> $resourcegroup="packtade"
    PS C:\>
    PS C:\> # Get the Azure Storage account context
    PS C:\> $storagecontext = (Get-AzStorageAccount -ResourceGroupName $resourcegroup -Name $storageaccountname).Context;
    PS C:\>
    PS C:\> #Create a new container
    PS C:\> New-AzStorageContainer -Name $containername -Context $storagecontext


      Blob End Point: https://packtadestorage.blob.core.windows.net/

    Name            PublicAccess        LastModified
    ----            ------------        ------------
    logfiles        Off                 3/29/2020 6:02:33 AM +00:00

    PS C:\>
    ```

    Figure 1.7 – Creating a new storage container

2. Execute the following commands to upload a text file to an existing container:

    ```
    #upload single file to container
    Set-AzStorageBlobContent -File "C:\ADECookbook\Chapter1\
    Logfiles\Logfile1.txt" -Context $storagecontext -Blob
    logfile1.txt -Container $containername
    ```

You should get a similar output to that shown in the following screenshot:

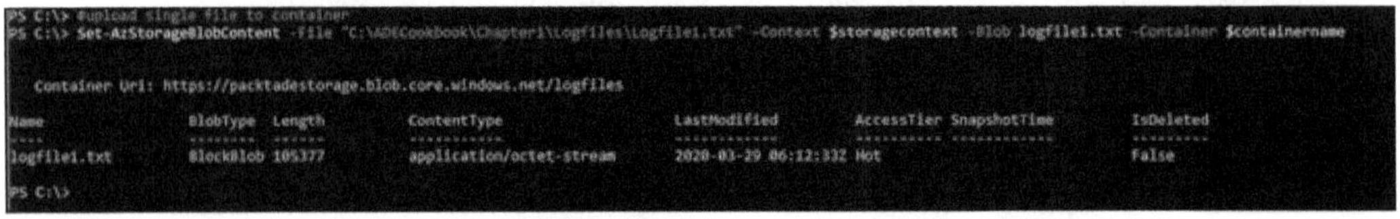

Figure 1.8 – Uploading a file to a storage container

3. Execute the following commands to upload all the files in a directory to an Azure container:

```
#get files to be uploaded from the directory
$files = Get-ChildItem -Path "C:\ADECookbook\Chapter1\
Logfiles";
#iterate through each file int the folder and upload it
to the azure container
foreach($file in $files){
    Set-AzStorageBlobContent -File $file.FullName
-Context $storagecontext -Blob $file.BaseName -Container
$containername -Force
}
```

You should get a similar output to that shown in the following screenshot:

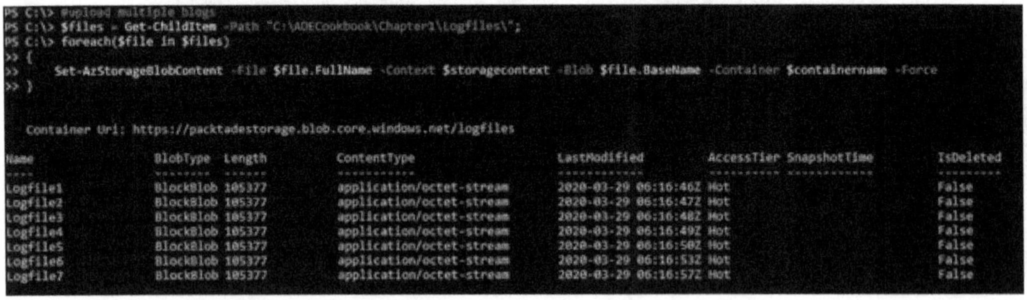

Figure 1.9 – Uploading multiple files to a storage container

How it works...

The storage container is created using the New-AzStorageContainer command. It takes two parameters – container name and storage context. The storage context can be set using the Get-AzStorageAccount command context property.

To upload files to the container, we used the `Set-AzStorageBlobContent` command. The command requires storage context, a file path to be uploaded, and the container name. To upload multiple files, we can iterate through the folder and upload each file using the `Set-AzStorageBlobContent` command.

Managing blobs in Azure Storage using PowerShell

In this recipe, we'll learn to perform various management tasks on an Azure blob. We'll copy blobs between containers, list all blobs in a container, modify a blob access tier, download blobs from Microsoft Azure to a local system, and delete a blob from Azure Storage.

Getting ready

Before you start, perform the following steps:

1. Make sure you have an existing Azure storage account. If not, create one by following the *Provisioning an Azure storage account using PowerShell* recipe.

2. Make sure you have an existing Azure storage container. If not, create one by following the *Creating containers and uploading files to Azure Blob storage using PowerShell* recipe.

3. Log in to your Azure subscription in PowerShell. To log in, run the `Connect-AzAccount` command in a new PowerShell window and follow the instructions.

How to do it...

Let's begin by copying blobs between containers.

Copying blobs between containers

Perform the following steps:

1. Execute the following commands to create a new container in an Azure storage account:

```
#set the parameter values
$storageaccountname="packtadestorage"
$resourcegroup="packtade"
$sourcecontainername="logfiles"
$destcontainername="textfiles"
```

```
#Get storage account context
$storagecontext = (Get-AzStorageAccount -ResourceGroupName
$resourcegroup -Name $storageaccountname).Context
# create the container
$destcontainer = New-AzStorageContainer -Name
$destcontainername -Context $storagecontext
```

You should get an output similar to the following screenshot:

```
PS C:\> # create the destination container
PS C:\> New-AzStorageContainer -Name $destcontainername -Context $storagecontext


   Blob End Point: https://packtadestorage.blob.core.windows.net/

Name                    PublicAccess              LastModified
----                    ------------              ------------
textfiles               Off                       3/30/2020 3:50:28 AM +00:00
```

Figure 1.10 – Creating a new storage container

2. Execute the following command to copy the `Logfile1` blob from the source container to the destination container:

```
#copy a single blob from one container to another
Start-CopyAzureStorageBlob -SrcBlob "Logfile1"
-SrcContainer $sourcecontainername -DestContainer
$destcontainername -Context $storagecontext -DestContext
$storagecontext
```

You should get an output similar to the following screenshot:

Figure 1.11 – Copying a blob from one storage container to another

3. Execute the following command to copy all the blobs from the source container to the destination container:

```
# copy all blobs in new container
Get-AzStorageBlob -Container $sourcecontainername
-Context $storagecontext | Start-CopyAzureStorageBlob
-DestContainer $destcontainername -DestContext
$storagecontext -force
```

You should get an output similar to the following screenshot:

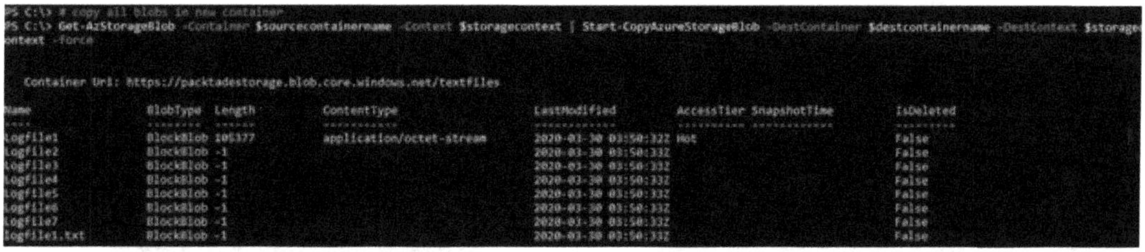

Figure 1.12 – Copying all blobs from one storage container to another

Listing blobs in an Azure storage container

Execute the following command to list the blobs from the destination container:

```
# list the blobs in the destination container
(Get-AzStorageContainer -Name $destcontainername -Context
$storagecontext).CloudBlobContainer.ListBlobs()
```

You should get an output similar to the following screenshot:

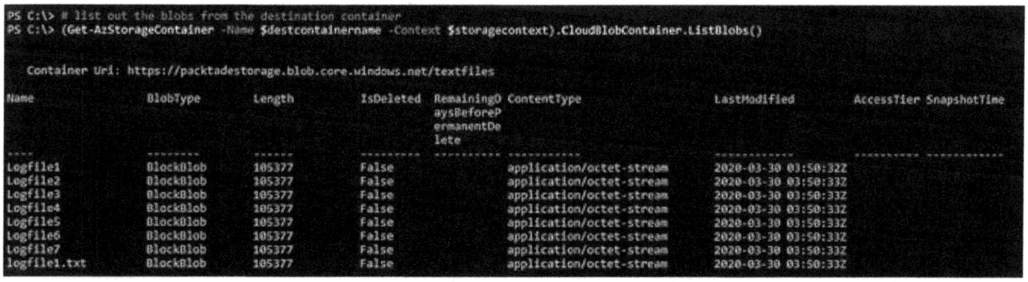

Figure 1.13 – Listing blobs in a storage container

Modifying a blob access tier

Perform the following steps:

1. Execute the following commands to change the access tier of a blob:

```
# Get the blob reference
$blob = Get-AzStorageBlob -Blob *Logfile4* -Container
$sourcecontainername -Context $storagecontext
#Get current access tier
$blob
#change access tier to cool
```

```
$blob.ICloudBlob.SetStandardBlobTier("Cool")
#Get the modified access tier
$blob
```

You should get an output similar to the following screenshot:

Figure 1.14 – Modifying the blob access tier

2. Execute the following commands to change the access tier of all the blobs in the container:

```
#get blob reference
$blobs = Get-AzStorageBlob -Container $destcontainername
-Context $storagecontext
#change the access tier of all the blobs in the container
$blobs.icloudblob.setstandardblobtier("Cool")
#verify the access tier
$blobs
```

You should get an output similar to the following screenshot:

Figure 1.15 – Modifying the blob access tier of all blobs in a storage container

Downloading a blob

Execute the following commands to download a blob from Azure Storage to your local computer:

```
#get the storage context
$storagecontext = (Get-AzStorageAccount -ResourceGroupName
$resourcegroup -Name $storageaccountname).Context
#download the blob
Get-AzStorageBlobContent -Blob "Logfile1" -Container
$sourcecontainername -Destination C:\ADECookbook\Chapter1\
Logfiles\ -Context $storagecontext -Force
```

Deleting a blob

Execute the following command to remove/delete a blob:

```
#get the storage context
$storagecontext = (Get-AzStorageAccount -ResourceGroupName
$resourcegroup -Name $storageaccountname).Context
Remove-AzStorageBlob -Blob "Logfile2" -Container
$sourcecontainername -Context $storagecontext
```

How it works...

Copying blobs across containers in the same storage account or in a different storage account can be done easily by the PowerShell Start-CopyAzureStorageBlob command. The command takes the source blob, destination blob, the source and destination containers, and the source and destination storage accounts as parameters. To copy all blobs in a container, we can run Get-AzStorageBlob to get all the blobs in the container and pipe it to the Start-CopyAzureStorageBlob command.

A blob access tier can be modified by first getting the reference to the blob object using Get-AzStorageBlob and then modifying the access tier using the setstandardblobtier property. There are three access tiers – Hot, Cool, and Archive:

- The Hot tier is suitable for files that are accessed frequently. It has a higher storage cost and low access cost.

- The Cool tier is suitable for infrequently accessed files and has a lower access cost and a lower storage cost.

- The Archive tier, as the name suggests, is used for long-term archival and should be used for files that are seldom required. It has the highest access cost and the lowest storage cost.

To download a blob from Azure to a local system, we use Get-AzStorageBlobContent. The command accepts the blob name, container name, local file path, and the storage context.

To delete a blob, run Remove-AzStorageBlob. Provide the blob name, container, and the storage context.

Managing an Azure blob snapshot in Azure Storage using PowerShell

An Azure blob snapshot is a point-in-time copy of a blob. A snapshot can be used as a blob backup. In this recipe, we'll learn to create, list, promote, and delete an Azure blob snapshot.

Getting ready

Before you start, perform the following steps:

1. Make sure you have an existing Azure storage account. If not, create one by following the *Provisioning an Azure storage account using PowerShell* recipe.

2. Make sure you have an existing Azure storage container. If not, create one by following the *Creating containers and uploading files to Azure Blob storage using PowerShell* recipe.

3. Make sure you have existing blobs/files in an Azure storage container. If not, you can upload blobs according to the previous recipe.

4. Log in to your Azure subscription in PowerShell. To log in, run the Connect-AzAccount command in a new PowerShell window and follow the instructions.

How to do it...

Let's begin by creating a blob snapshot.

Creating a blob snapshot

Execute the following commands to create a new blob snapshot:

```
#set the parameter values
$storageaccountname="packtadestorage"
$resourcegroup="packtade"
$sourcecontainername ="logfiles"
#get the storage context
$storagecontext = (Get-AzStorageAccount -ResourceGroupName
$resourcegroup -Name $storageaccountname).Context
#get blob context
$blob = Get-AzStorageBlob -Blob *Logfile5* -Container
$sourcecontainername -Context $storagecontext
#create blob snapshot
$blob.ICloudBlob.CreateSnapshot()
```

You should get an output similar to the following screenshot:

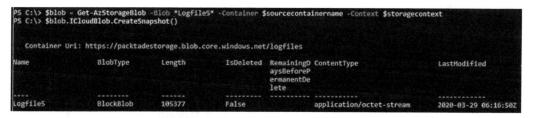

Figure 1.16 – Creating a blob snapshot

Listing snapshots

Execute the following commands to list all snapshots for the blobs in a given container:

```
# get blobs in a container
$blobs = Get-AzStorageBlob -Container $sourcecontainername
-Context $storagecontext
#list snapshots
$blobs | Where-Object{$_.ICloudBlob.IsSnapshot -eq $true}
```

You should get an output similar to the following screenshot:

```
PS C:\> $blobs | Where-Object{$_.ICloudBlob.IsSnapshot -eq $true}


    Container Uri: https://packtadstorage.blob.core.windows.net/logfiles

Name                    BlobType  Length      ContentType                LastModified              AccessTier SnapshotTime
----                    --------  ------      -----------                ------------              ---------- ------------
Logfile5                BlockBlob 105377      application/octet-stream    2020-03-29 06:16:50Z Cool         2020-03-30 06:14:0
```

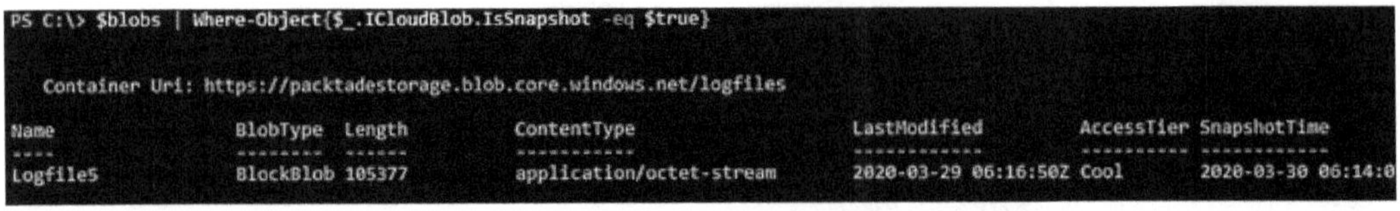

Figure 1.17 – Listing blob snapshots

Promoting a snapshot

Promoting a snapshot refers to replacing the existing blob with the blob snapshot. To promote a snapshot, copy the snapshot onto the original blob.

Execute the following commands to promote the snapshot:

```
#get reference of original blob
$blob = Get-AzStorageBlob -Blob *Logfile5* -Container
$sourcecontainername -Context $storagecontext
$originalblob = $blob | Where-Object{$_.ICloudBlob.IsSnapshot
-eq $false}
#get reference of the blob snapshot
$blobsnapshot = $blob | Where-Object{$_.ICloudBlob.IsSnapshot
-eq $true}
#overwrite the original blob with the snapshot
Start-CopyAzureStorageBlob -CloudBlob $blobsnapshot.
ICloudBlob -DestCloudBlob $originalblob.ICloudBlob -Context
$storagecontext -Force
```

You should get an output similar to the following screenshot:

```
PS C:\> $blob = Get-AzStorageBlob -Blob *Logfile5* -Container $sourcecontainername -Context $storagecontext
PS C:\> $originalblob = $blob | Where-Object{$_.ICloudBlob.IsSnapshot -eq $false}
PS C:\> $blobsnapshot = $blob | Where-Object{$_.ICloudBlob.IsSnapshot -eq $true}
PS C:\> Start-CopyAzureStorageBlob -CloudBlob $blobsnapshot.ICloudBlob -DestCloudBlob $originalblob.ICloudBlob -Context $stora


    Container Uri: https://packtadstorage.blob.core.windows.net/logfiles

Name                    BlobType  Length      ContentType                LastModified              AccessTier SnapshotTime
----                    --------  ------      -----------                ------------              ---------- ------------
Logfile5                BlockBlob 105377      application/octet-stream    2020-03-30 08:01:34Z Hot
```

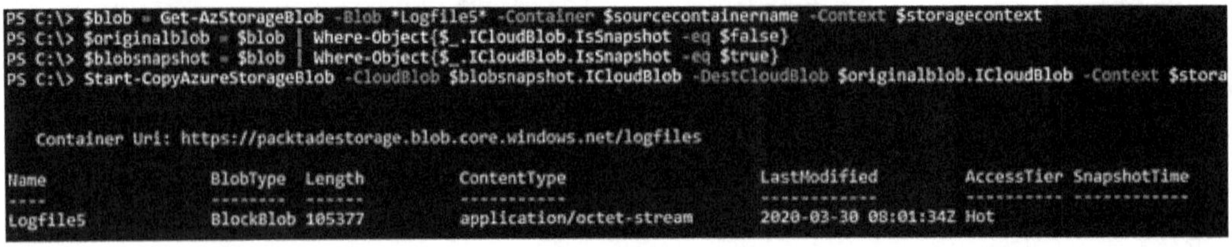

Figure 1.18 – Promoting a blob snapshot

Deleting a blob snapshot

Execute the following command to delete a blob snapshot:

```
Remove-AzStorageBlob -CloudBlob $originalblob.ICloudBlob
-DeleteSnapshot -Context $storagecontext
```

How it works...

A blob snapshot is a point-in-time copy of the blob. A snapshot can't be modified and can be used to revert to the point in time if required. A blob snapshot is charged separately and therefore should be created as per the requirement. To create a blob snapshot, get the blob reference using the `Get-AzStorageBlob` command and then create the snapshot using the `CreateSnapshot` function.

To promote a snapshot, we can overwrite the original blob by the snapshot using the `Start-CopyAzureStorageBlob` command.

To delete a snapshot, run the `Remove-AzStorageBlob` command and specify the `DeleteSnapshot` switch.

Configuring blob life cycle management for blob objects using the Azure portal

Azure Storage provides different blob access tiers such as `Hot`, `Cool`, and `Archive`. Each access tier has a different storage and data transfer cost. Applying a proper life cycle rule to move a blob among different access tiers helps optimize the cost. In this recipe, we'll learn to apply a life cycle rule to a blob using the Azure portal.

Getting ready

Before you start, perform the following steps:

1. Make sure you have an existing Azure storage account. If not, create one by following the *Provisioning an Azure storage account using PowerShell* recipe.
2. Make sure you have an existing Azure storage container. If not, create one by following the *Creating containers and uploading files to Azure Blob storage using PowerShell* recipe.

3. Make sure you have existing blobs/files in an Azure storage container. If not, you can upload blobs in accordance with the previous recipe. Then, log in to the Azure portal at, `https://portal.azure.com`.

How to do it...

Follow the given steps to configure a blob life cycle:

1. In Azure Portal, find and open the Azure storage case. In our case, it's `packtadestorage`.

2. In the **packtadestorage** window, search and select **Lifecycle Management** under **Blob service**, as shown in the following screenshot:

Figure 1.19 – Opening Lifecycle Management

3. In the **Add a rule** page, under the **Action set** tab, provide the life cycle
    configuration. A life cycle defines when to move a blob from a hot to a cool access
    tier, when to move a blob from a cool to a storage access tier, and when to delete
    the blob. These configurations are optional. We can also provide when to delete a
    blob snapshot. Then, click the **Next : Filter set >** button to continue, as shown in the
    following screenshot:

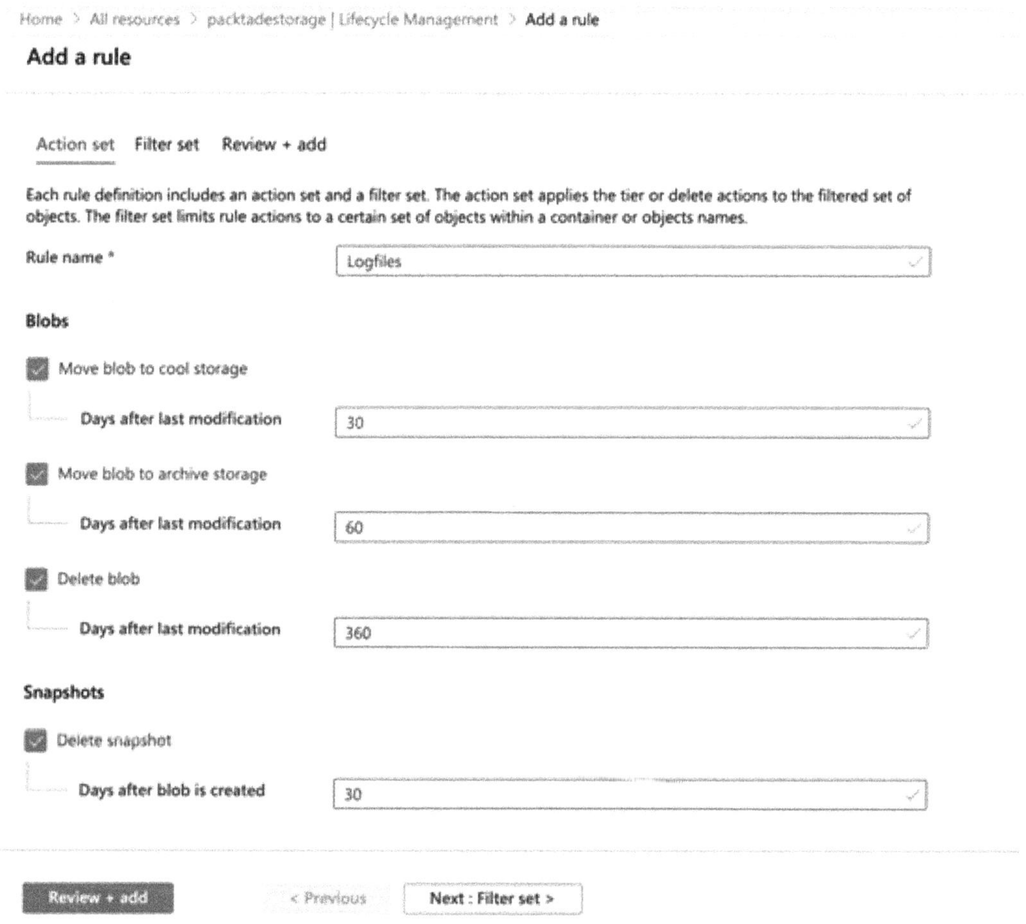

Figure 1.20 – Lifecycle Management – Action set

4. In the **Filter set** tab, specify the container or the virtual folder to apply the life cycle rule. Then, click the **Review + add** button to continue:

Figure 1.21 – Lifecycle Management – Filter set

5. In the **Review + add** page, review the configuration and click **Create** to create the rule:

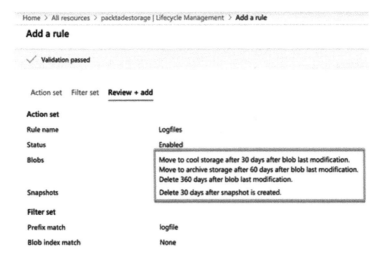

Figure 1.22 – Lifecycle Management – Review + add

How it works...

A blob life cycle management rule helps in managing storage costs by modifying the access tier of blobs as per the specified rule. Consider a log processing application that reads the log file from Azure storage, analyzes it, and saves the result in a database. As the log file is read and processed, it may not be needed any further. Therefore, moving it to a cool access tier from a hot access tier will save on storage costs.

Blob life cycle management helps in automating the access tier modification as per the application requirement and is therefore a must-have for any storage-based application.

Configuring a firewall for an Azure storage account using the Azure portal

Storage account access can be restricted to an IP or range of IPs by whitelisting the allowed IPs in the storage account firewall. In this recipe, we'll learn to restrict access to an Azure storage account using a firewall.

Getting ready

Before you start, perform the following steps:

1. Open a web browser and go to the Azure portal at `https://portal.azure.com`.

2. Make sure you have an existing storage account. If not, create one using the *Provisioning an Azure storage account using the Azure portal* recipe.

How to do it...

To provide access to an IP or range of IPs, follow the given steps:

1. In the Azure portal, locate and open the Azure storage account. In our case, the storage account is `packtadestorage`.

2. On the storage account page, under the **Settings** section, locate and select **Firewalls and virtual networks**.

3. As the `packtadestorage` account was created with public access, (check the *Provisioning an Azure storage account using the Azure portal* recipe), it can be accessed from all networks. To allow access from an IP or an IP range, select the **Selected networks** option on the storage account, from the **Firewalls and virtual networks** page:

Figure 1.23 – Azure Storage – Firewalls and virtual networks

4. In the **Selected networks** option, scroll down to the **Firewall** section. To give access to your machine only, select the **Add your client IP address** option. To give access to a different IP or range of IPs, type in the IPs under the **Address range** section:

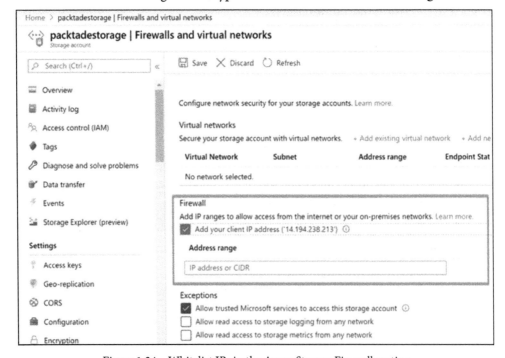

Figure 1.24 – Whitelist IPs in the Azure Storage Firewall section

5. To access the storage accounts from Azure services such as Azure Data Factory and Azure Functions, check **Allow trusted Microsoft services to access this storage account** under the **Exceptions** heading.

6. Click **Save** to save the configuration changes.

How it works...

Firewall settings are used to restrict access to an Azure storage account to an IP or range of IPs. Even if a storage account is public, it will only be accessible to the whitelisted IPs defined in the firewall configuration.

Configuring virtual networks for an Azure storage account using the Azure portal

A storage account can be public, accessible to everyone, public with access to an IP or range of IPs, or private with access to selected virtual networks. In this recipe, we'll learn to restrict access to an Azure storage account to a virtual network.

Getting ready

Before you start, perform the following steps:

1. Open a web browser and go to the Azure portal at `https://portal.azure.com`.

2. Make sure you have an existing storage account. If not, create one using the *Provisioning an Azure storage account using the Azure portal* recipe.

How to do it...

To restrict access to a virtual network, follow the given steps:

1. In the Azure portal, locate and open the storage account. In our case, it's `packtadestorage`. On the storage account page, under the **Settings** section, locate and select **Firewalls and virtual networks | Selected networks**:

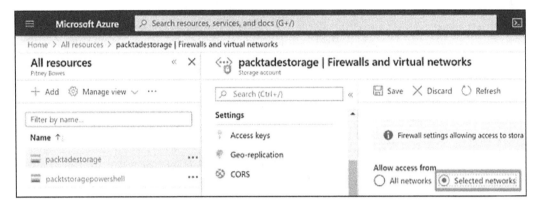

Figure 1.25 – Azure Storage – Selected networks

2. Under the **Virtual networks** section, select **+ Add new virtual network**:

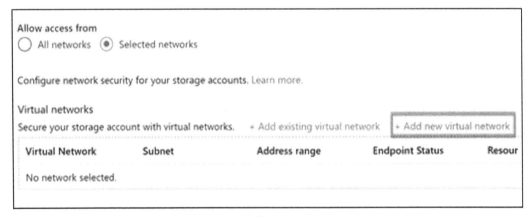

Figure 1.26 – Adding a virtual network

3. In the **Create virtual network** blade, provide the virtual network name, address space, and subnet address range. The remainder of the configuration values are pre-filled:

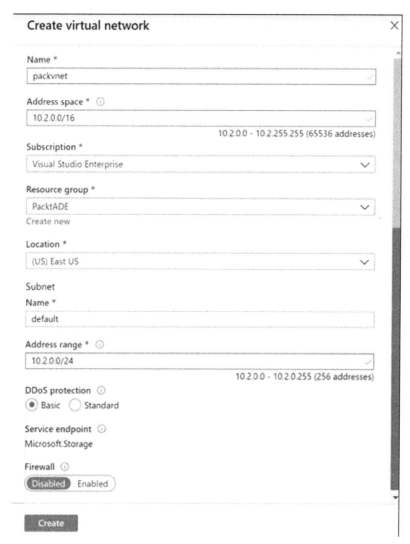

Figure 1.27 – Creating a new virtual network

4. Click **Create** to create the virtual network. The virtual network is created and listed under the **Virtual Network** section, as shown in the following screenshot:

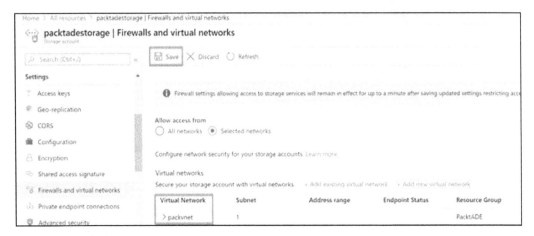

Figure 1.28 – Saving a virtual network configuration

5. Click **Save** to save the configuration changes.

How it works...

We first created an Azure virtual network and then added it to the Azure storage account. Creating the Azure virtual network from the storage account page automatically fills in the resource group, location, and subscription information. The virtual network and the storage account should be in same location.

The address space specifies the number of IP addresses in the given virtual network.

We also need to define the subnet within the virtual network the storage account will belong to. We can create a custom subnet. In our case, for the sake of simplicity, we have used the default subnet.

This allows the storage account to only be accessed by resources that belong to the given virtual network. The storage account is inaccessible to any network other than the specified virtual network.

Configuring a firewall for an Azure storage account using PowerShell

In this recipe, we'll enable firewall rules for an Azure storage account using PowerShell.

Getting ready

Before you start, perform the following steps:

1. Make sure you have an existing Azure storage account. If not, create one by following the *Provisioning an Azure storage account using PowerShell* recipe.

2. Log in to your Azure subscription in PowerShell. To log in, run the `Connect-AzAccount` command in a new PowerShell window and follow the instructions.

How to do it...

The steps for this recipe are as follows:

1. Execute the following command to deny access from all networks:

```
Update-AzStorageAccountNetworkRuleSet -ResourceGroupName
packtADE -Name packtadestorage -DefaultAction Deny
```

You should get a similar output to that shown in the following screenshot:

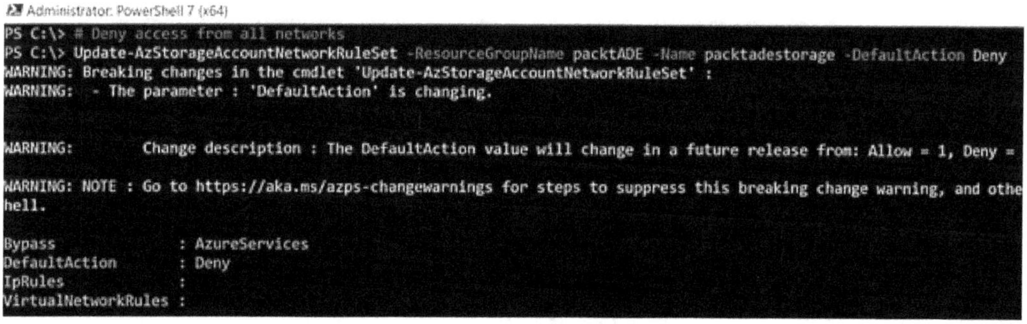

Figure 1.29 – Denying access to all networks

2. Execute the following commands to add a firewall rule for the client IP address:

```
#get client IP Address
$mypublicIP = (Invoke-WebRequest -uri "http://ifconfig.
me/ip").Content
#Add client IP address firewall rule
Add-AzStorageAccountNetworkRule -ResourceGroupName
packtADE -AccountName packtadestorage -IPAddressOrRange
$mypublicIP
```

You should get a similar output to that shown in the following screenshot:

```
PS C:\> #get client IP Address
PS C:\>
PS C:\> $mypublicIP = (Invoke-WebRequest -uri "http://ifconfig.me/ip").Content
PS C:\>
PS C:\> #Add client IP address firewall rule
PS C:\>
PS C:\> Add-AzStorageAccountNetworkRule -ResourceGroupName packtADE -AccountName packtadestorage -IPAddressOrRange $mypublicIP

Action IPAddressOrRange
------ ----------------
 Allow 182.156.106.173
```

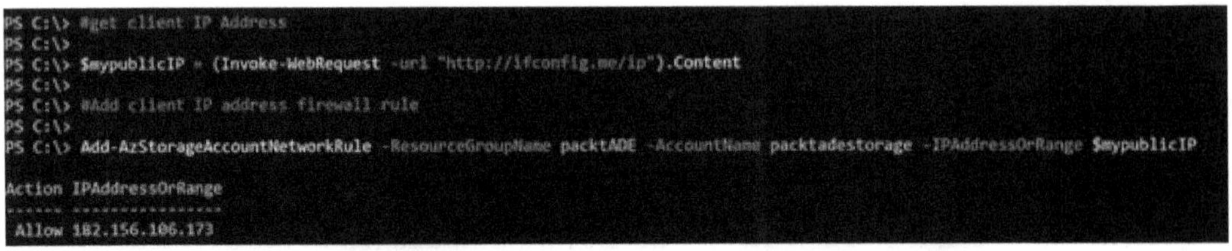

Figure 1.30 – Adding a host public IP to a firewall

3. Execute the following command to whitelist a custom IP to access the storage account:

```
#whitelist a single IP
Add-AzStorageAccountNetworkRule -ResourceGroupName
packtADE -AccountName packtadestorage -IPAddressOrRange
"20.24.29.30"
```

You should get a similar output to that shown in the following screenshot:

```
PS C:\>
PS C:\> # whitelist a single IP
>>
>> Add-AzStorageAccountNetworkRule -ResourceGroupName packtADE -AccountName packtadestorage -IPAddressOrRange "20.24.29.30"
>>
>>

Action IPAddressOrRange
------ ----------------
 Allow 182.156.106.173
 Allow 20.24.29.30
```

Figure 1.31 – Adding a custom IP to a firewall

4. Execute the following command to whitelist a custom IP range to access the storage account:

```
#whitelist range of IPs
Add-AzStorageAccountNetworkRule -ResourceGroupName
packtADE -AccountName packtadestorage -IPAddressOrRange
"20.24.0.0/24"
```

You should get a similar output to that shown in the following screenshot:

```
PS C:\> #whitelist range of IPs
PS C:\>
PS C:\> Add-AzStorageAccountNetworkRule -ResourceGroupName packtADE -AccountName packtadestorage -IPAddressOrRange "20.24.0.0/24"

Action IPAddressOrRange
------ ----------------
 Allow 182.156.106.173
 Allow 20.24.29.30
 Allow 20.24.0.0/24
```

Figure 1.32 – Adding a custom IP range to a firewall

5. Execute the following command to get all the existing firewall rules:

```
(Get-AzStorageAccountNetworkRuleSet -ResourceGroupName
packtADE -Name packtadestorage).IpRules
```

You should get a similar output to that shown in the following screenshot:

```
PS C:\> (Get-AzStorageAccountNetworkRuleSet -ResourceGroupName packtADE -Name packtadestorage).IpRules

Action IPAddressOrRange
------ ----------------
 Allow 182.156.106.173
 Allow 20.24.29.30
 Allow 20.24.0.0/24
```

Figure 1.33 – Listing existing firewall rules

6. Execute the following commands to remove the firewall rules:

```
#Remove the client IP from the firewall rule
Remove-AzStorageAccountNetworkRule -ResourceGroupName
packtADE -Name packtadestorage -IPAddressOrRange
$mypublicIP
#Remove the single IP from the firewall rule
Remove-AzStorageAccountNetworkRule -ResourceGroupName
packtADE -Name packtadestorage -IPAddressOrRange
"20.24.29.30"
#Remove the IP range from the firewall rule
Remove-AzStorageAccountNetworkRule -ResourceGroupName
packtADE -Name packtadestorage -IPAddressOrRange
"20.24.0.0/24"
```

You should get the following output:

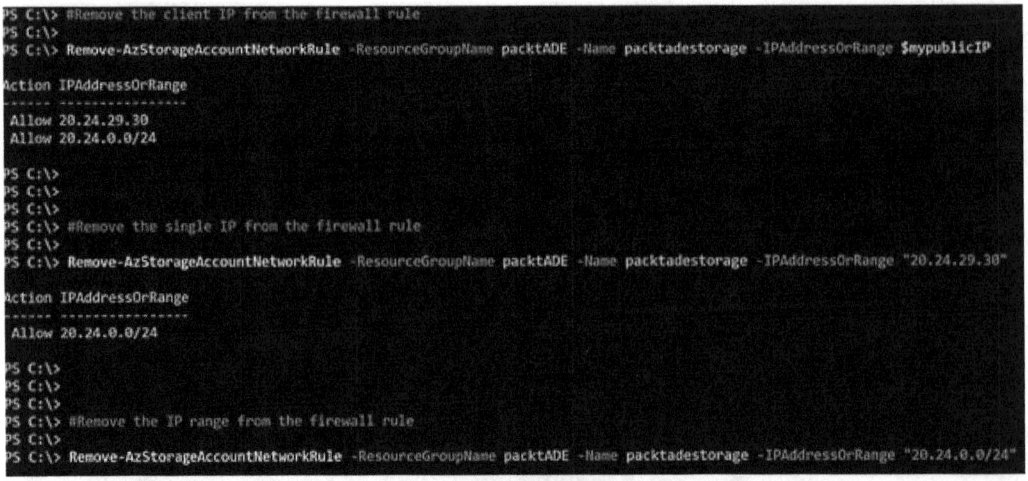

Figure 1.34 – Removing firewall rules

7. Execute the following command to allow access to all networks:

```
Update-AzStorageAccountNetworkRuleSet -ResourceGroupName
packtADE -Name packtadestorage -DefaultAction Allow
```

You should get the following output:

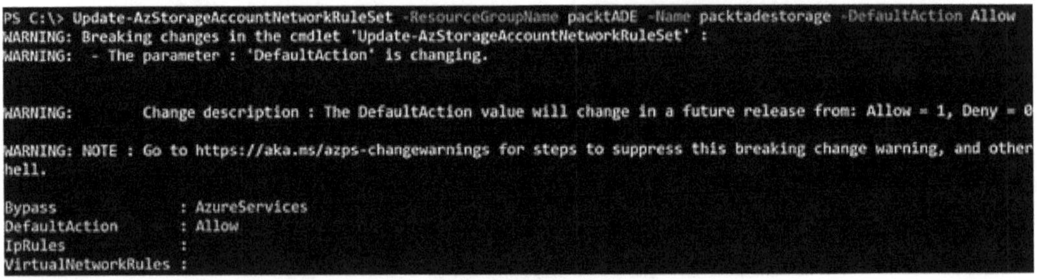

Figure 1.35 – Allowing access to all networks

How it works...

To whitelist an IP or range of IPs, we first need to modify the storage account to use selected networks instead of all networks. This is done by means of the Update-AzStorageAccountNetworkRuleSet command.

We can then whitelist an IP or range of IPs using the Add-AzStorageAccountNetworkRule command. We provide the resource group name, storage account name, and the IP or range of IPs to whitelist.

We can get the list of existing rules using the `Get-AzStorageAccountNet workRuleSet` command by providing the resource group and the storage account name as the parameter.

We can remove the IPs from the firewall using the `Remove-AzStorageAccount NetworkRule` command by providing the resource group name, storage account name, and the IP or the IP range to remove.

Configuring virtual networks for an Azure storage account using PowerShell

In this recipe, we'll learn to limit access of an Azure storage account to a particular virtual network using PowerShell.

Getting ready

Before you start, perform the following steps:

1. Make sure you have an existing Azure storage account. If not, create one by following the *Provisioning an Azure storage account using PowerShell* recipe.

2. Log in to your Azure subscription in PowerShell. To log in, run the `Connect-AzAccount` command in a new PowerShell window and follow the instructions.

How to do it...

The steps for this recipe are as follows:

1. Execute the following command to disable all networks for an Azure storage account:

```
$resourcegroup = "packtADE"
$location="eastus"
Update-AzStorageAccountNetworkRuleSet -ResourceGroupName
$resourcegroup -Name packtadestorage -DefaultAction Deny
```

2. Execute the following command to create a new virtual network (if you want to provide access to an existing virtual network, you can skip this step):

```
#create a new virtual network
New-AzVirtualNetwork -Name packtVnet -ResourceGroupName
$resourcegroup -Location $location -AddressPrefix
"10.1.0.0/16"
```

You should get a similar output to that shown in the following screenshot:

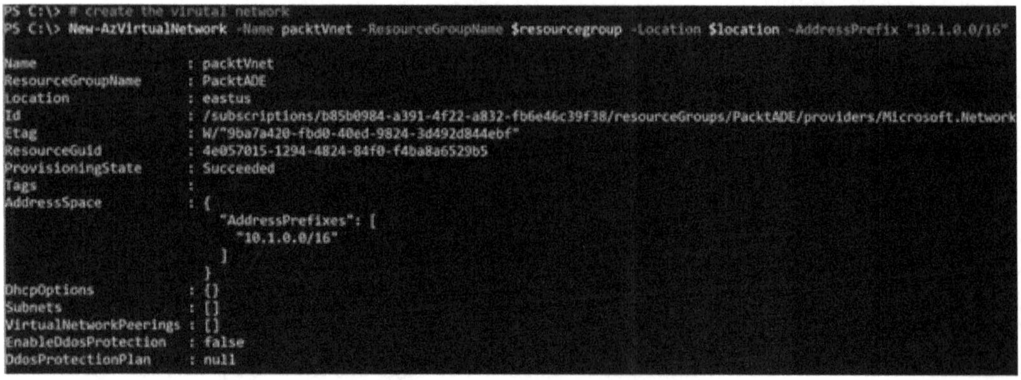

Figure 1.36 – Creating a new virtual network

3. Execute the following command to get the virtual network details:

```
$vnet = Get-AzVirtualNetwork -Name packtvnet
-ResourceGroupName packtade
```

4. Execute the following command to create a new subnet:

```
Add-AzVirtualNetworkSubnetConfig -Name default
-VirtualNetwork $vnet -AddressPrefix "10.1.0.0/24"
-ServiceEndpoint "Microsoft.Storage"
```

You should get a similar output to that shown in the following screenshot:

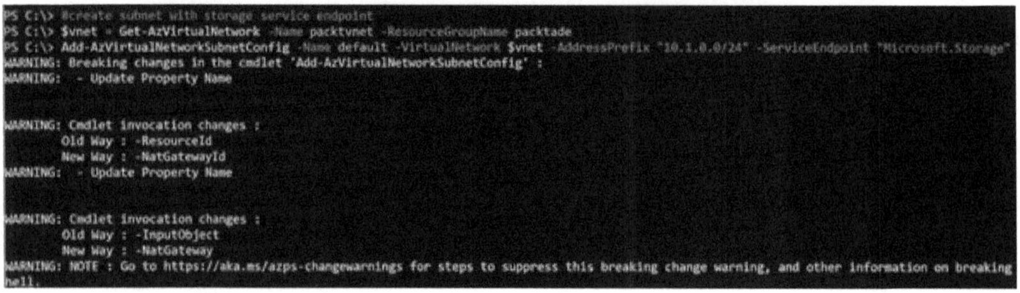

Figure 1.37 – Creating a new subnet

5. Execute the following command to confirm and apply the preceding changes:

```
$vnet | Set-AzVirtualNetwork
```

You should get a similar output to that shown in the following screenshot:

```
PS C:\> $vnet | Set-AzVirtualNetwork

Name               : packtVnet
ResourceGroupName  : packtade
Location           : eastus
Id                 : /subscriptions/b85b0984-a391-4f22-a832-fb6e46c39f38/resourceGroups/pack
Etag               : W/"f3794889-8515-4c87-945b-72e26f1b4f17"
ResourceGuid       : 4e057015-1294-4824-84f0-f4ba8a6529b5
ProvisioningState  : Succeeded
Tags               :
AddressSpace       : {
                       "AddressPrefixes": [
                         "10.1.0.0/16"
                       ]
                     }
DhcpOptions        : {
                       "DnsServers": []
                     }
Subnets            : [
                       {
                         "Delegations": [],
                         "Name": "default",
                         "Etag": "W/\"f3794889-8515-4c87-945b-72e26f1b4f17\"",
                         "Id": "/subscriptions/b85b0984-a391-4f22-a832-fb6e46c39f38/resource
                     /subnets/default",
```

Figure 1.38 – Saving virtual network settings

6. Execute the following command to fetch the virtual network and subnet details into a variable:

```
$vnet = Get-AzVirtualNetwork -Name packtvnet
-ResourceGroupName $resourcegroup
$subnet = Get-AzVirtualNetworkSubnetConfig -Name default
-VirtualNetwork $vnet
```

7. Execute the following command to add the virtual network to an Azure storage account:

```
Add-AzStorageAccountNetworkRule -ResourceGroupName
$resourcegroup -Name packtadestorage
-VirtualNetworkResourceId $subnet.Id
```

You should get a similar output to that shown in the following screenshot:

```
PS C:\> Add-AzStorageAccountNetworkRule -ResourceGroupName packtade -Name packtadestorage -VirtualNetworkResourceId $subnet.Id

Action  VirtualNetworkResourceId                                                                                                              State
------  ------------------------                                                                                                              -----
Allow   /subscriptions/b85b0984-a391-4f22-fb6e46c39f38/resourceGroups/packtade/providers/Microsoft.Network/virtualNetworks/packtVnet/subnets/default  Succeeded

PS C:\>
```

Figure 1.39 – Adding a virtual network to Azure Storage

8. Execute the following command to fetch all virtual network rules on an Azure storage account:

```
(Get-AzStorageAccountNetworkRuleSet -ResourceGroupName
$resourcegroup -Name packtadestorage).VirtualNetworkRules
```

You should get the following output:

Figure 1.40 – Listing existing virtual network rules

9. Execute the following command to remove the virtual network rule from an Azure storage account:

```
Remove-AzStorageAccountNetworkRule -ResourceGroupName
$resourcegroup -Name packtadestorage
-VirtualNetworkResourceId $subnet.id
```

10. Execute the following command to remove the virtual network:

```
Remove-AzVirtualNetwork -Name packtvnet
-ResourceGroupName $resourcegroup -Force
```

11. Execute the following command to provide access to all networks:

```
Update-AzStorageAccountNetworkRuleSet -ResourceGroupName
$resourcegroup -Name packtadestorage -DefaultAction Allow
```

How it works...

We can restrict access to an Azure storage account to a virtual network. To do this, we first run the Update-AzStorageAccountNetworkRuleSet command to deny access to an Azure storage account to all networks.

We then use the New-AzVirtualNetwork command to create a new virtual network. A virtual network requires an address prefix that defines the IP ranges in the network. Any resource that belongs to that virtual network will get an IP or IPs from the IP range defined in the virtual network address prefix.

The next step is to create a subnet in the virtual network using the `Add-AzVirtualNetworkSubnetConfig` command. A subnet also has an address prefix. However, this should be within the virtual network address prefix. The subnet is created with the `Microsoft.Storage` service endpoint as we need to add this to the Azure storage account.

In the next step, we run the `Add-AzStorageAccountNetworkRule` command to add the subnet to the virtual network rule in the storage account. The command expects `virtualnetworkresourceId`, which is the subnet ID. We can get the subnet ID by running the `Get-AzStorageAccountNetworkSubnetConfig` command and specifying the subnet name and the virtual network object.

Once we add the virtual network rule, storage account access is limited to the resources that belong to the given virtual network.

Creating an alert to monitor an Azure storage account

We can create an alert on multiple available metrics to monitor an Azure storage account. To create an alert, we need to define the trigger condition and the action to be performed when the alert is triggered. In this recipe, we'll create an alert to send an email if the used capacity metrics for an Azure storage account exceed 5 MB. The used capacity threshold of 5 MB is not a standard and is deliberately kept low to explain the alert functionality.

Getting ready

Before you start, perform the following steps:

1. Open a web browser and log in to the Azure portal at `https://portal.azure.com`.
2. Make sure you have an existing storage account. If not, create one using the *Provisioning an Azure storage account using the Azure portal* recipe.

How to do it...

Follow the given steps to create an alert:

1. In the Azure portal, locate and open the storage account. In our case, the storage account is `packtadestorage`. On the storage account page, search and open **Alerts** under the **Monitoring** section:

Figure 1.41 – Selecting Alerts

2. On the **Alerts** page, click on **New alert rule**:

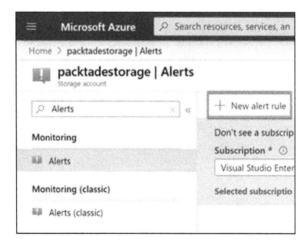

Figure 1.42 – Adding a new alert

3. On the **Alerts | Create rule** page, observe that the storage account is listed by default under the **RESOURCE** section. You can add multiple storage accounts in the same alert. Under the **CONDITION** section, click **Add**:

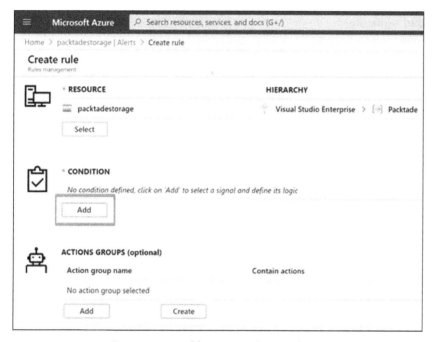

Figure 1.43 – Adding a new alert condition

4. On the **Configure signal logic** page, select **Used capacity** under **Signal name**:

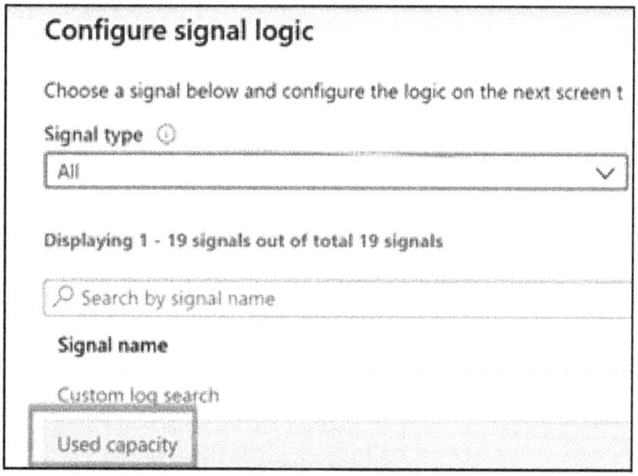

Figure 1.44 – Configuring signal logic

5. On the **Configure signal logic** page, under **Alert logic**, set **Operator** as **Greater than**, **Aggregation type** as **Average**, and configure the threshold to 5 MB. We need to provide the value in bytes:

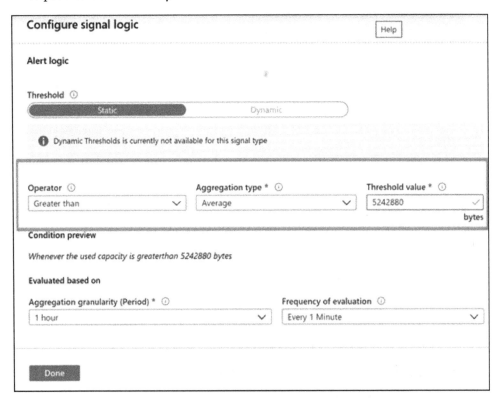

Figure 1.45 – Configuring alert logic

6. Click **Done** to configure the trigger. The condition is added, and we'll be taken back to the **Configure alerts rule** page:

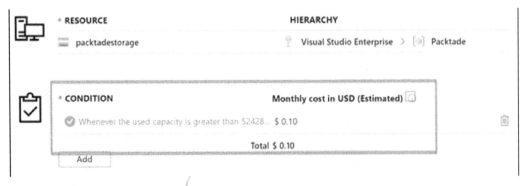

Figure 1.46 – Viewing a new alert condition

7. The next step is to add an action to perform when the alert condition is reached. On the **Configure alerts rule** page, under the **ACTIONS GROUPS** section, click **Create**:

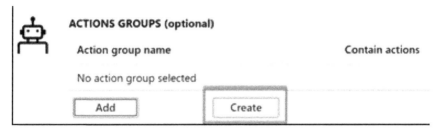

Figure 1.47 – Creating a new alert action group

8. On the **Add action group** page, provide the action group name, short name, and resource group. Under the **Actions** section, provide the action name and action type:

Figure 1.48 – Adding a new alert action group

9. As we set **Action Type** as **Email/SMS/Push/Voice**, a new blade opens. In the
    Email/SMS/Push/Voice blade, specify the email name and click **OK**:

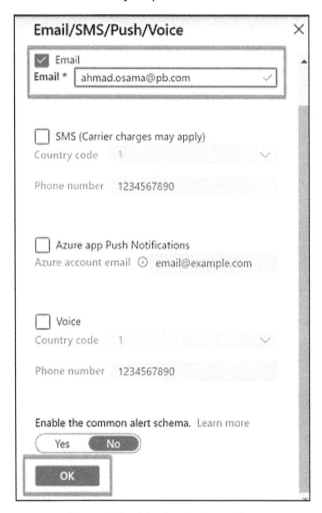

Figure 1.49 – Selecting the alert action

10. We are taken back to the **Add action group** page. On the **Add action group** page,
    click **OK** to save the action settings. We are then taken back to the **Create rule** page.
    The Email action is listed under the **Action Groups** section.

11. The next step is to define the alert rule name, description, and severity:

Figure 1.50 – Create alert rule

12. Click the **Create alert rule** button to create the alert.

13. The next step is to trigger the alert. To do that, upload the `~/Chapter1/Logfiles/Logfile15.txt` file to the Azure storage account following the steps mentioned in the *Creating containers and uploading files to Azure Blob storage using PowerShell* recipe. The triggered alerts are listed on the **Alerts** page, as shown in the following screenshot:

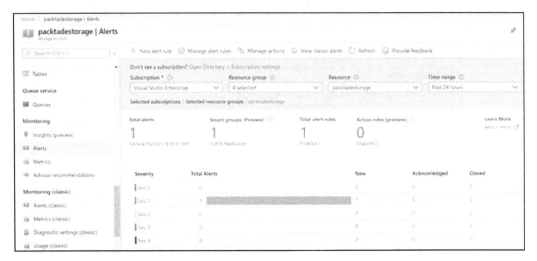

Figure 1.51 – Viewing alerts

An email is sent to the email ID specified in the email action group. The email appears as shown in the following snapshot:

Figure 1.52 – Alert email

How it works...

Setting up an alert is easy. At first, we need to define the alert condition (trigger or signal). An alert condition defines the metrics and the threshold that, when breached, the alert is to be triggered. We can define more than one condition on multiple metrics for one alert.

We then need to define the action to be performed when the alert condition is reached. We can define more than one action for an alert. In our example, in addition to sending an email when the used capacity is more than 5 MB, we can configure Azure Automation to delete the old blobs/files so as to maintain the Azure storage capacity within 5 MB.

There are other signals such as transactions, Ingress, Egress, Availability, Success Server Latency, and Success E2E Latency on which alerts can be defined. Detailed information on monitoring Azure storage is available at `https://docs.microsoft.com/en-us/azure/storage/common/storage-monitoring-diagnosing-troubleshooting`.

Securing an Azure storage account with SAS using PowerShell

A **shared access signature** (**SAS**) provides more granular access to blobs by specifying an expiry limit, specific permissions, and IPs.

Using SAS, we can specify different permissions to users or application on different blobs based on the requirement. For example, if an application needs to read one file/blob from a container, instead of providing access to all the files in the container, we can use SAS to provide read access on the required blob.

In this recipe we'll learn to create and use SAS to access blobs.

Getting ready

Before you start, go through the following steps:

- Make sure you have an existing Azure storage account. If not, create one by following the *Provisioning an Azure storage account using PowerShell* recipe.

- Make sure you have an existing Azure storage container. If not, create one by following the *Creating containers and uploading files to Azure Blob storage using PowerShell* recipe.

- Make sure you have existing blobs/files in an Azure storage container. If not, you can upload blobs by following the previous recipe.

- Log in to your Azure subscription in PowerShell. To log in, run the `Connect-AzAccount` command in a new PowerShell window and follow the instructions.

How to do it...

Let's begin by securing blobs using SAS.

Securing blobs using SAS

Perform the following steps:

1. Execute the following command in the PowerShell window to get the storage context:

```
$resourcegroup = "packtade"
$storageaccount = "packtadestorage"
#get storage context
$storagecontext = (Get-AzStorageAccount
-ResourceGroupName $resourcegroup -Name $storageaccount).
Context
```

2. Execute the following commands to get the SAS token for the `logfile1.txt` blob in the logfiles container with list and read permissions:

```
#set the token expiry time
$starttime = Get-Date
$endtime = $starttime.AddDays(1)
# get the SAS token into a variable
$sastoken = New-AzStorageBlobSASToken -Container
"logfiles" -Blob "logfile1.txt" -Permission lr -StartTime
$starttime -ExpiryTime $endtime -Context $storagecontext
# view the SAS token.
$sastoken
```

3. Execute the following commands to list the blob using the SAS token:

```
#get storage account context using the SAS token
$ctx = New-AzStorageContext -StorageAccountName
$storageaccount -SasToken $sastoken
```

```
#list the blob details
Get-AzStorageBlob -blob "logfile1.txt" -Container
"logfiles" -Context $ctx
```

You should get an output as shown in the following screenshot:

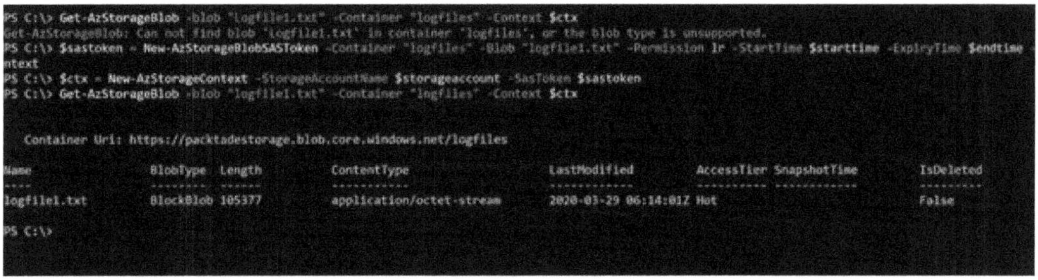

Figure 1.53 – Listing blobs using SAS

4. Execute the following command to write data to `logfile1.txt`:

```
Set-AzStorageBlobContent -File C:\ADECookbook\Chapter1\
Logfiles\Logfile1.txt -Container logfiles -Context $ctx
```

You should get an output as shown in the following screenshot:

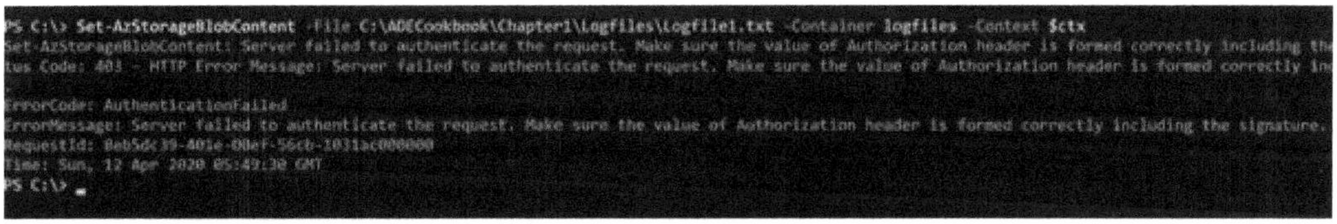

Figure 1.54 – Uploading a blob using SAS

The write fails as the SAS token was created with list and read access.

Securing a container with SAS

Perform the following steps:

1. Execute the following command to create a container stored access policy:

```
$resourcegroup = "packtade"
$storageaccount = "packtadestorage"
#get storage context
$storagecontext = (Get-AzStorageAccount
-ResourceGroupName $resourcegroup -Name $storageaccount).
Context
$starttime = Get-Date
```

```
$endtime = $starttime.AddDays(1)
New-AzStorageContainerStoredAccessPolicy -Container
logfiles -Policy writepolicy -Permission lw -StartTime
$starttime -ExpiryTime $endtime -Context $storagecontext
```

2. Execute the following command to create the SAS token:

```
# get the SAS token
$sastoken = New-AzStorageContainerSASToken -Name logfiles
-Policy writepolicy -Context $storagecontext
```

3. Execute the following commands to list all the blobs in the container using the SAS token:

```
#get the storage context with SAS token
$ctx = New-AzStorageContext -StorageAccountName
$storageaccount -SasToken $sastoken
#list blobs using SAS token
Get-AzStorageBlob -Container logfiles -Context $ctx
```

How it works...

To generate a shared access token for a blob, use the `New-AzStorageBlobSASToken` command. We need to provide the blob name, container name, permission (l=list, r=read, and w=write), and storage context to generate a SAS token. We can additionally secure the token by providing IPs that can access the blob.

We then use the SAS token to get the storage context using the `New-AzStorageContext` command. We use the storage context to access the blobs using the `Get-AzStorageBlob` command. Observe that we can only list and read blobs and can't write to the blob as the SAS token doesn't have write permission.

To generate a shared access token for a container, we first create an access policy for the container using the `New-AzStorageContainerStoredAccessPolicy` command. The access policy specifies the start and expiry time, permission, and the IPs. We then generate the SAS token by passing the access policy name to the `New-AzStorageContainerSASToken` command.

We can now access the container and the blobs using the SAS token.

2
Working with Relational Databases in Azure

Microsoft Azure provides Azure SQL Database, PostgreSQL, and MySQL as Database-as-a-Service offerings. We can create an instance of these databases without worrying about the installation, administration, infrastructure, and upgrades.

Needless to say that we can install any of the available **Relational Database Management System** (**RDBMS**) databases, such as Oracle or DB2, on an Azure **virtual machine** (**VM**).

In a data pipeline, we can use any of the RDBMS databases as either a source or a destination. In this chapter, we'll learn how to provision, connect to, manage, maintain, and secure these databases.

We'll cover the following recipes in this chapter:

- Provisioning and connecting to an Azure SQL database using PowerShell
- Provisioning and connecting to an Azure PostgreSQL database using the Azure CLI
- Provisioning and connecting to an Azure MySQL database using the Azure CLI

- Implementing active geo-replication for an Azure SQL database using PowerShell
- Implementing an auto-failover group for an Azure SQL database using PowerShell
- Implementing vertical scaling for an Azure SQL database using PowerShell
- Implementing an Azure SQL database elastic pool using PowerShell
- Monitoring an Azure SQL database using the Azure portal

Provisioning and connecting to an Azure SQL database using PowerShell

In this recipe, we'll learn how to create and connect to an Azure SQL Database instance. Azure SQL Database comes in three failovers: **standalone Azure SQL Database**, **Azure SQL Database elastic pools**, and **managed instances**. In this recipe, we'll create a standalone Azure SQL database.

Getting ready

In a new PowerShell window, execute the Connect-AzAccount command to log in to your Microsoft Azure account.

How to do it...

Let's begin by provisioning Azure SQL Database.

Provisioning Azure SQL Database

The steps are as follows:

1. Execute the following PowerShell command to create a new resource group:

   ```
   New-AzResourceGroup -Name packtade -Location "central us"
   ```

2. Execute the following query to create a new Azure SQL server:

   ```
   #create credential object for the Azure SQL Server admin
   credential
   $sqladminpassword = ConvertTo-SecureString 'Sql@
   Server@1234' -AsPlainText -Force
   $sqladmincredential = New-Object System.Management.
   Automation.PSCredential ('sqladmin', $sqladminpassword)
   ```

```
# create the azure sql server
New-AzSqlServer -ServerName azadesqlserver
-SqlAdministratorCredentials $sqladmincredential
-Location "central us" -ResourceGroupName packtade
```

You should get a similar output as shown in the following screenshot:

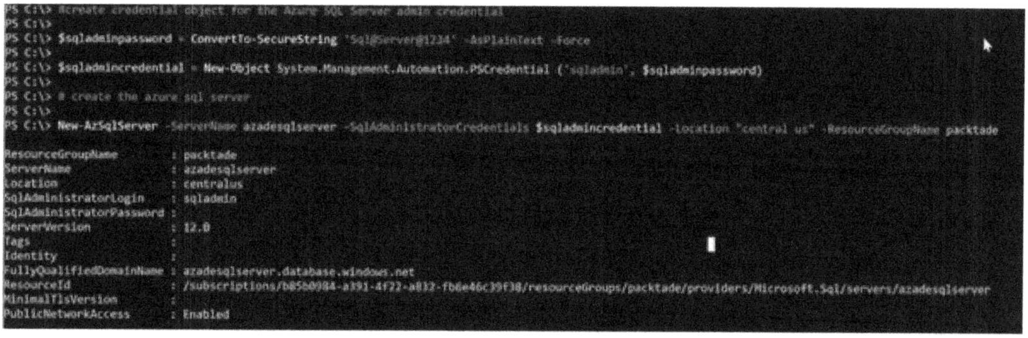

Figure 2.1 – Creating a new Azure SQL server

3. Execute the following query to create a new Azure SQL database:

```
New-AzSqlDatabase -DatabaseName azadesqldb -Edition basic
-ServerName azadesqlserver -ResourceGroupName packtade
```

You should get an output as shown in the following screenshot:

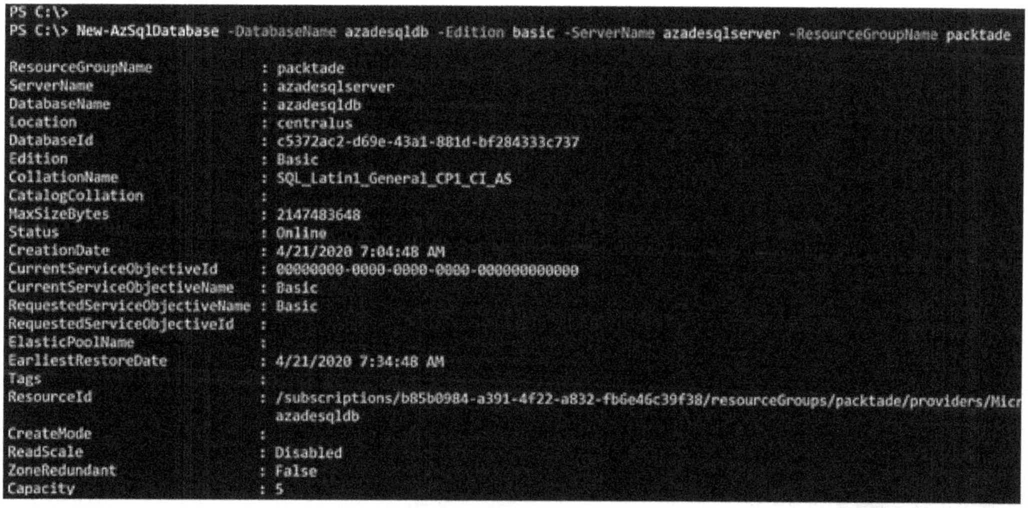

Figure 2.2 – Creating a new Azure SQL database

Connecting to an Azure SQL database

To connect to an Azure SQL database, let's first whitelist the IP in the Azure SQL Server firewall:

1. Execute the following command to whitelist the public IP of the machine to connect to an Azure SQL database. (This recipe assumes that you are connecting from your local system. To connect from a system other than your local system, change the IP in the following command.) Execute the following command in the PowerShell window to whitelist the machine's public IP in the Azure SQL Server firewall:

    ```
    $clientip = (Invoke-RestMethod -Uri https://ipinfo.io/
    json).ip
    New-AzSqlServerFirewallRule -FirewallRuleName "home"
    -StartIpAddress $clientip -EndIpAddress $clientip
    -ServerName azadesqlserver -ResourceGroupName packtade
    ```

    You will get an output similar to the one shown in the following screenshot:

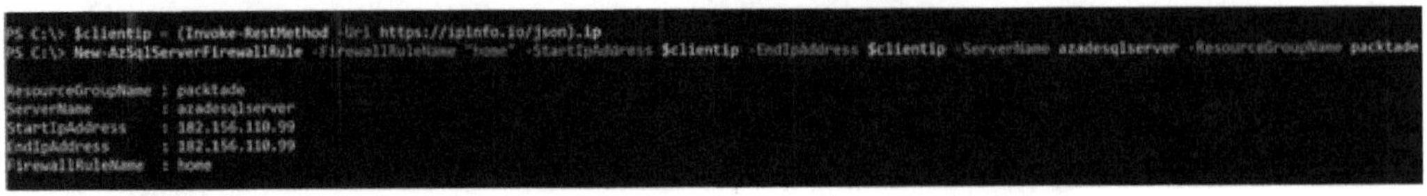

Figure 2.3 – Creating a new Azure SQL Server firewall rule

2. Execute the following command to connect to an Azure SQL database from SQLCMD (SQLCMD comes with the SQL Server installation, or you can download the **SQLCMD** utility from https://docs.microsoft.com/en-us/sql/tools/sqlcmd-utility?view=sql-server-ver15):

    ```
    sqlcmd -S "azadesqlserver.database.windows.net" -U
    sqladmin -P "Sql@Server@1234" -d azadesqldb
    ```

    Here's the output:

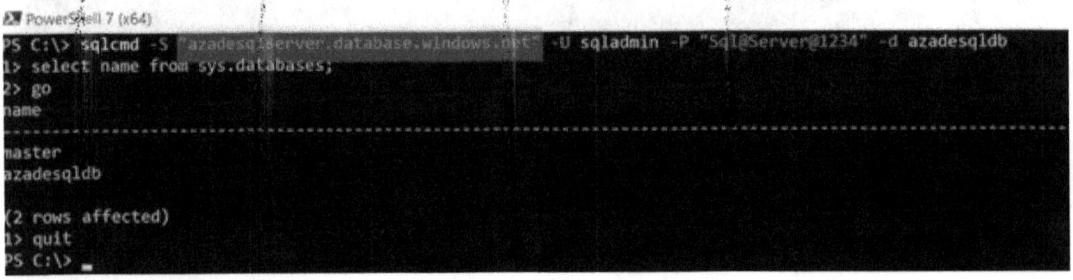

Figure 2.4 – Connecting to an Azure SQL database

How it works...

We first execute the `New-AzSQLServer` command to provision a new Azure SQL server. The command accepts the server name, location, resource group, and login credentials. An Azure SQL server, unlike an on-premises SQL server, is not a physical machine or VM that is accessible to customers.

We then execute the `New-AzSQLDatabase` command to create an Azure SQL database. This command accepts the database name, the Azure SQL server name, the resource group, and the edition. There are multiple SQL database editions to choose from based on the application workload. However, for the sake of this demo, we will create a basic edition.

To connect to an Azure SQL database, we first need to whitelist the machine's IP in the Azure SQL Server firewall. Only whitelisted IPs are allowed to connect to the database. To whitelist the client's public IP, we use the `New-AzSQLServerfirewallrule` command. This command accepts the server name, resource group, and start and end IPs. We can either whitelist a single IP or a range of IPs.

We can connect to an Azure SQL database from SQL Server Management Studio, SQLCMD, or a programming language using the appropriate SQL Server drivers. When connecting to an Azure SQL database, we need to specify the server name as `azuresqlservername.database.windows.net`, and then specify the Azure SQL database to connect to.

Provisioning and connecting to an Azure PostgreSQL database using the Azure CLI

Azure Database for PostgreSQL is a Database-as-a-Service offering for the PostgreSQL database. In this recipe, we'll learn how to provision an Azure database for PostgreSQL and connect to it.

Getting ready

We'll be using the Azure CLI for this recipe. Open a new Command Prompt or PowerShell window, and run `az login` to log in to the Azure CLI.

How to do it...

Let's begin with provisioning a new Azure PostgreSQL server.

Provisioning a new Azure PostrgreSQL server

The steps are as follows:

1. Execute the following Azure CLI command to create a new resource group:

   ```
   az group create --name rgpgressql --location eastus
   ```

2. Execute the following command to create an Azure server for PostgreSQL:

   ```
   az postgres server create --resource-group rgpgressql
   --name adepgresqlserver  --location eastus --admin-user
   pgadmin --admin-password postgre@SQL@1234 --sku-name B_
   Gen5_1
   ```

   > **Note**
   > It may take 10–15 minutes for the server to be created.

3. Execute the following command to whitelist the IP in the PostgreSQL server firewall:

   ```
   $clientip = (Invoke-RestMethod -Uri https://ipinfo.io/
   json).ip
   ```

   ```
   az postgres server firewall-rule create --resource-
   group rgpgressql --server adepgresqlserver --name hostip
   --start-ip-address $clientip --end-ip-address $clientip
   ```

Connecting to an Azure PostgreSQL server

We can connect to an Azure PostgreSQL server using `psql` or `pgadmin` (a GUI tool for PostgreSQL management), or from any programming language using a relevant driver.

To connect from `psql`, execute the following command in a Command Prompt or PowerShell window:

```
PS C:\Program Files\PostgreSQL\12\bin> .\psql.exe
--host=adepgresqlserver.postgres.database.azure.com --port=5432
--username=pgadmin@adepgresqlserver --dbname=postgres
```

Provide the password and you'll be connected. You should get an output similar to the one shown in the following screenshot:

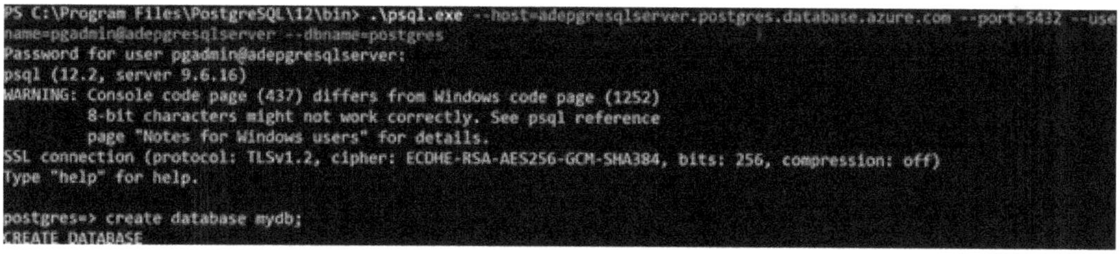

Figure 2.5 – Connecting to PostgreSQL

How it works...

To provision a new Azure PostgreSQL server, execute the following Azure CLI command – `az postgres server create`. We need to specify the server name, resource group, administrator username and password, location, and SKU name parameters. As of now, there are three different SKUs:

- `B_Gen5_1` is the basic and smallest SKU, up to 2 vCores.

- `GP_Gen5_32` is the general-purpose SKU, up to 64 vCores.

- `MO_Gen5_2` is the memory-optimized SKU, with 32 memory-optimized vCores.

> **Note**
>
> For more information on the pricing tiers, visit `https://docs.microsoft.com/en-us/azure/postgresql/concepts-pricing-tiers`.

To connect to the PostgreSQL server, we first need to whitelist the IP in the server firewall. To do that, we run the `az postgres server firewall-rule create` Azure CLI command.

We need to provide the firewall rule name, server name, resource group, and start and end IP.

Once the firewall rule is created, the PostgreSQL server can be accessed by any of the utilities, such as `psql` or `pgadmin`, or from a programming language. To connect to the server, provide the host or server name as `<postgresql server name>.postgres.database.azure.com` and the port as `5432`. We also need to provide the username and password. If you are connecting for the first time, provide the database name as `postgres`.

Provisioning and connecting to an Azure MySQL database using the Azure CLI

Azure Database for MySQL is a Database-as-a-Service offering for the MySQL database. In this recipe, we'll learn how to provision an Azure database for MySQL and connect to it.

Getting ready

We'll be using the Azure CLI for this recipe. Open a new Command Prompt or PowerShell window, and run az login to log in to the Azure CLI.

How it works...

Let's see how to provision the Azure MySQL server.

Provisioning the Azure MySQL server

The steps are as follows:

1. Execute the following command to create a new resource group:

```
az group create --name rgmysql --location eastus
```

2. Execute the following command to provision a new Azure MySQL server:

```
az mysql server create --resource-group rgmysql --name
ademysqlserver  --location eastus --admin-user dbadmin
--admin-password mySQL@1234 --sku-name B_Gen5_1
```

You should get an output as shown in the following screenshot:

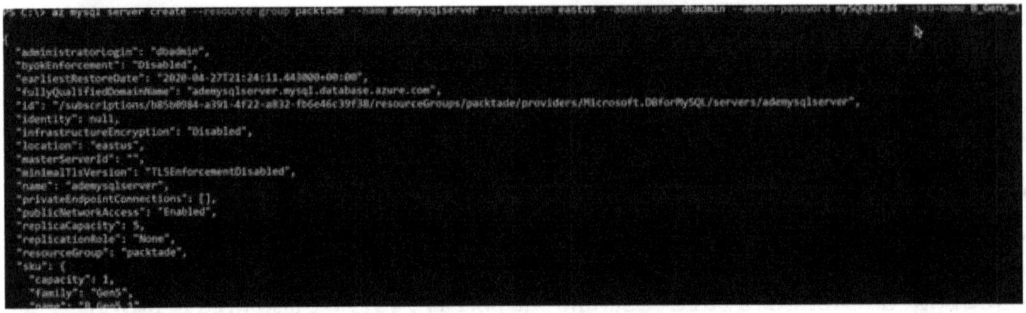

Figure 2.6 – Creating an Azure MySQL server

Connecting to Azure MySQL Server

The steps are as follows:

1. Execute the following command to whitelist your public IP in the Azure MySQL Server firewall:

```
$clientip = (Invoke-RestMethod -Uri https://ipinfo.io/
json).ip
az mysql server firewall-rule create --resource-group
rgmysql --server ademysqlserver --name clientIP --start-
ip-address $clientip --end-ip-address $clientip
```

You should get an output as shown in the following screenshot:

Figure 2.7 – Creating a firewall rule for the Azure MySQL Server

2. We can connect to the Azure MySQL server using the MySQL shell or the MySQL workbench, or from any programming language. To connect from the MySQL shell, execute the following command in a PowerShell window:

```
.\mysqlsh.exe -h ademysqlserver.mysql.database.azure.com
-u dbadmin@ademysqlserver -p
```

Here's the output:

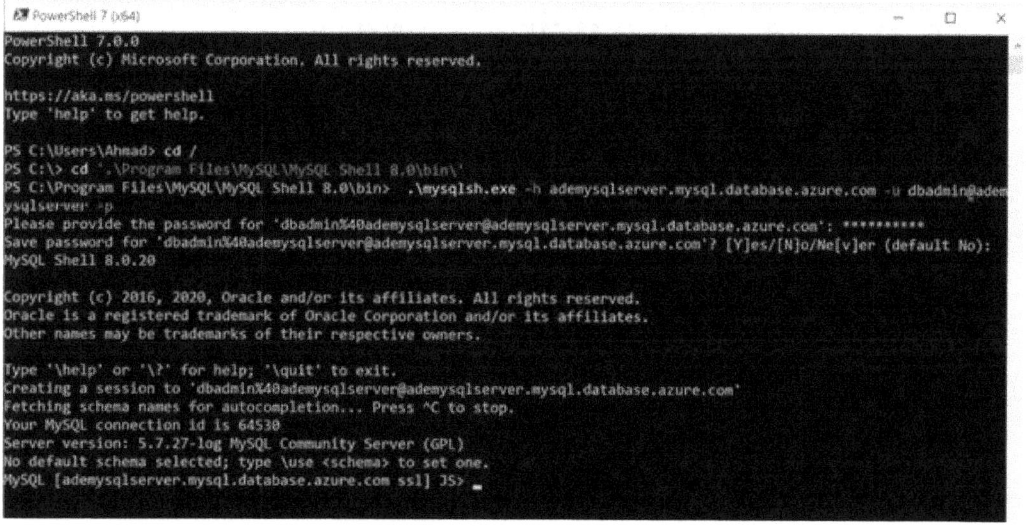

Figure 2.8 – Connecting to the Azure MySQL server

How it works...

To provision a new Azure MySQL server, execute the following Azure CLI command `- az mysql server create`. We need to specify the server name, resource group, administrator username and password, location, and SKU name parameters. As of now, there are three different SKUs:

- `B_Gen5_1` is the basic and smallest SKU, up to 2 vCores.

- `GP_Gen5_32` is the general-purpose SKU, up to 64 vCores.

- `MO_Gen5_2` is the memory-optimized SKU, with 32 memory-optimized vCores.

To connect to the MySQL server, we first need to whitelist the IP in the server firewall. To do that, we run the `az mysql server firewall-rule create` Azure CLI command.

We need to provide the firewall rule name, server name, resource group, and start and end IPs.

Once the firewall rule is created, the MySQL server can be accessed by any of the utilities, such as the MySQL command line or the MySQL workbench, or from a programming language. To connect to the server, provide the host or server name as `<mysql server name>.mysql.database.azure.com`. We also need to provide the username and password.

Implementing active geo-replication for an Azure SQL database using PowerShell

The active geo-replication feature allows you to create up to four readable secondaries of a primary Azure SQL database. Active geo-replication uses **SQL Server AlwaysOn** to asynchronously replicate transactions to the secondary databases. The secondary database can be in the same or a different region than the primary database.

Active geo-replication can be used for the following cases:

- To provide business continuity by failing over to the secondary database in case of a disaster. The failover is manual.

- To offload reads to the readable secondary.

- To migrate a database to a different server in another region.

In this recipe, we'll configure active geo-replication for an Azure SQL database and perform a manual failover.

Getting ready

In a new PowerShell window, execute the `Connect-AzAccount` command and follow the steps to log in to your Azure account.

You need an existing Azure SQL database for this recipe. If you don't have one, create an Azure SQL database by following the steps mentioned in the *Provisioning and connecting to an Azure SQL database using PowerShell* recipe.

How to do it...

First, let's create a readable secondary.

Creating a readable secondary

The steps are as follows:

1. Execute the following command to provision a new Azure SQL server to host the secondary replica:

    ```
    #create credential object for the Azure SQL Server admin
    credential
    $sqladminpassword = ConvertTo-SecureString 'Sql@
    Server@1234' -AsPlainText -Force
    $sqladmincredential = New-Object System.Management.
    Automation.PSCredential ('sqladmin', $sqladminpassword)
    New-AzSQLServer -ServerName azadesqlsecondary
    -SqlAdministratorCredentials $sqladmincredential
    -Location westus -ResourceGroupName packtade
    ```

2. Execute the following command to configure the geo-replication from the primary server to the secondary server:

    ```
    $primarydb = Get-AzSqlDatabase -DatabaseName azadesqldb
    -ServerName azadesqlserver -ResourceGroupName packtade
    $primarydb | New-AzSqlDatabaseSecondary
    -PartnerResourceGroupName packtade -PartnerServerName
    azadesqlsecondary -AllowConnections "All"
    ```

You should get an output as shown in the following screenshot:

PS C:\> $primarydb = Get-AzSqlDatabase -DatabaseName azadesqldb -ServerName azadesqlserver -ResourceGroupName packtade
PS C:\> $primarydb | New-AzSqlDatabaseSecondary -PartnerResourceGroupName packtade -PartnerServerName azadesqlsecondary -AllowConnections "All"
LinkId : 17c927ca-a2d7-4d8c-857a-34851ed460f0
ResourceGroupName : packtade
ServerName : azadesqlserver
DatabaseName : azadesqldb
Role : Primary
Location : Central US
PartnerResourceGroupName : packtade
PartnerServerName : azadesqlsecondary
PartnerDatabaseName : azadesqldb
PartnerRole : Secondary
PartnerLocation : West US
AllowConnections : All
ReplicationState : CATCH_UP
PercentComplete : 100
StartTime : 4/30/2020 8:26:27 PM

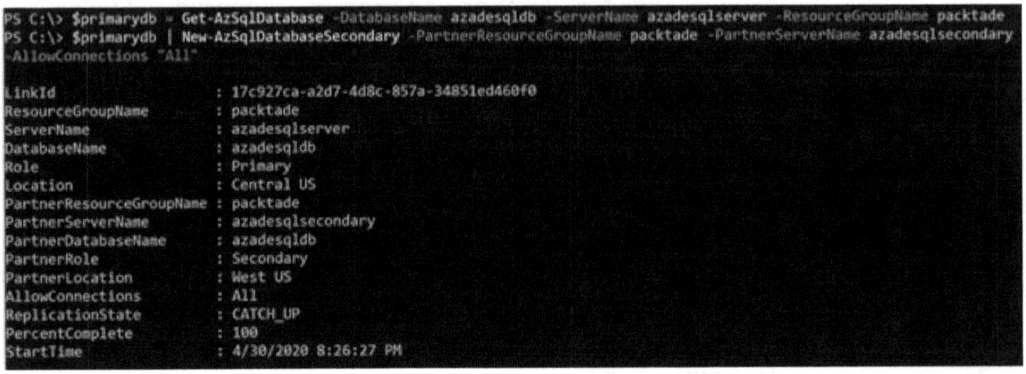

Figure 2.9 – Configuring geo-replication

Moreover, we can also check this on the Azure portal, as shown in the following screenshot:

Microsoft Azure
azadesqldb (azadesqlserver/azadesqldb) | Geo-Replication
Overview
Activity log
Tags
Diagnose and solve problems
Quick start
Query editor (preview)
Power Platform
Power BI (preview)
PowerApps (preview)
Flow (preview)
Settings
Configure
Geo-Replication
Server/Database
Failover policy
Status
Primary
Central US
azadesqlserver/azadesqldb
None
Online
Secondaries
West US
azadesqlsecondary/azadesqldb
Seeding 0%

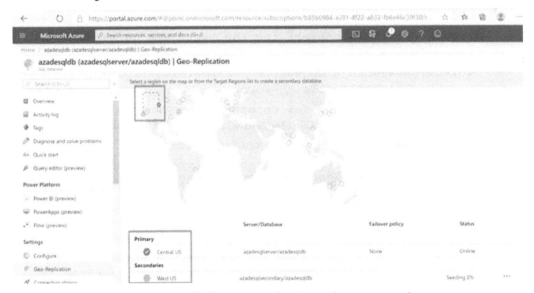

Figure 2.10 – Verifying geo-replication in the Azure portal

Performing manual failover to the secondary

The steps are as follows:

1. Execute the following command to manually failover to the secondary database:

    ```
    $secondarydb = Get-AzSqlDatabase -DatabaseName azadesqldb
    -ServerName azadesqlsecondary -ResourceGroupName packtade
    ```
    ```
    $secondarydb | Set-AzSqlDatabaseSecondary
    -PartnerResourceGroupName packtade -Failover
    ```

    The preceding command performs a planned failover without data loss. To perform a manual failover with data loss, use the `Allowdataloss` switch.

    If we check the Azure portal, we'll see that `azadesqlsecondary/azadesqldb` in **West US** is the primary database:

Figure 2.11 – Failing over to the secondary server

2. We can also get the active geo-replication information by executing the following command:

    ```
    Get-AzSqlDatabaseReplicationLink -DatabaseName azadesqldb
    -PartnerResourceGroupName packtade -PartnerServerName
    azadesqlsecondary -ServerName azadesqlserver
    -ResourceGroupName packtade
    ```

You should get an output as shown in the following screenshot:

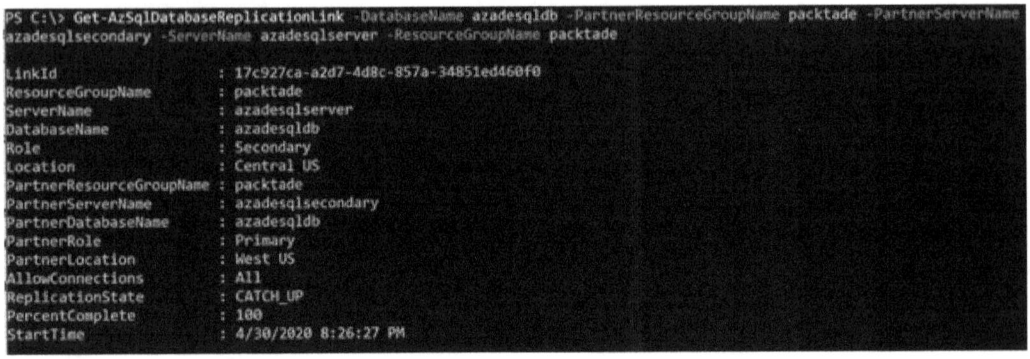

Figure 2.12 – Getting the geo-replication status

Removing active geo-replication

Execute the following command to remove the active geo-replication link between the primary and the secondary databases:

```
$primarydb = Get-AzSqlDatabase -DatabaseName azadesqldb
-ServerName azadesqlserver -ResourceGroupName packtade
```

```
$primarydb | Remove-AzSqlDatabaseSecondary
-PartnerResourceGroupName packtade -PartnerServerName
azadesqlsecondary
```

You should get an output as shown in the following screenshot:

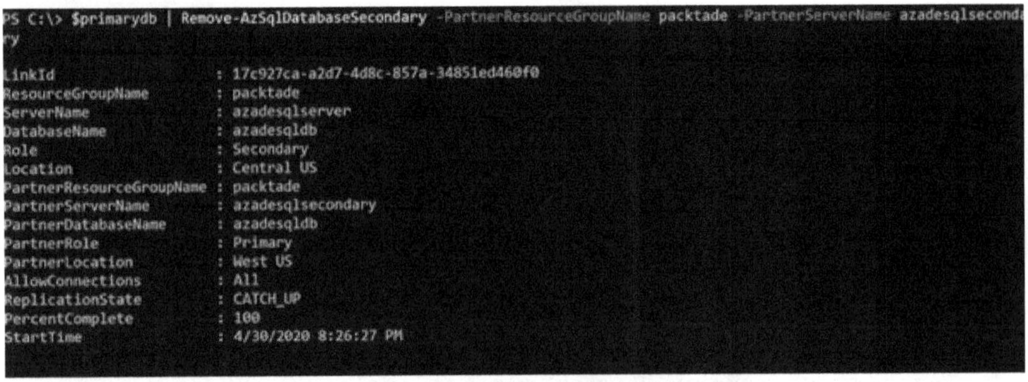

Figure 2.13 – Removing active geo-replication

How it works...

To configure active geo-replication, we use the `New-AzSqlDatabaseSecondary` command. This command expects the primary database name, server name, and resource group name, and the secondary resource group name, server name, and the `Allow connections` parameter. If we want a readable secondary, then we set `Allow connections` to `All`; otherwise, we set it to `No`.

The active geo-replication provides manual failover with and without data loss. To perform a manual failover, we use the `Set-AzSqlDatabaseSecondary` command. This command expects the secondary server name, database name, resource group name, a failover switch, and the `Allowdataloss` switch in case of failover with data loss.

To remove active geo-replication, we use the `Remove-AzSqlDatabaseSecondary` command. This command expects the secondary server name, secondary database name, and resource name to remove the replication link between the primary and the secondary database.

Removing active geo-replication doesn't remove the secondary database.

Implementing an auto-failover group for an Azure SQL database using PowerShell

An auto-failover group allows a group of databases to fail to a secondary server in another region in case the SQL database service in the primary region fails. Unlike active geo-replication, the secondary server should be in a different region than the primary. The secondary databases can be used to offload read workloads.

The failover can be manual or automatic.

In this recipe, we'll create an auto-failover group, add databases to the auto-failover group, and perform a manual failover to the secondary server.

Getting ready

In a new PowerShell window, execute the `Connect-AzAccount` command and follow the steps to log in to your Azure account.

You will need an existing Azure SQL database for this recipe. If you don't have one, create an Azure SQL database by following the steps mentioned in the *Provisioning and connecting to an Azure SQL database using PowerShell* recipe.

How to do it...

First, let's create an auto-failover group.

Creating an auto-failover group

The steps are as follows:

1. Execute the following PowerShell command to create a secondary server. The server should be in a different region than the primary server:

    ```
    $sqladminpassword = ConvertTo-SecureString 'Sql@
    Server@1234' -AsPlainText -Force
    ```

    ```
    $sqladmincredential = New-Object System.Management.
    Automation.PSCredential ('sqladmin', $sqladminpassword)
    ```

    ```
    New-AzSQLServer -ServerName azadesqlsecondary
    -SqlAdministratorCredentials $sqladmincredential
    -Location westus -ResourceGroupName packtade
    ```

2. Execute the following command to create the auto-failover group:

    ```
    New-AzSqlDatabaseFailoverGroup -ServerName azadesqlserver
    -FailoverGroupName adefg -PartnerResourceGroupName
    packtade -PartnerServerName azadesqlsecondary
    -FailoverPolicy Automatic -ResourceGroupName packtade
    ```

    You should get an output as shown in the following screenshot:

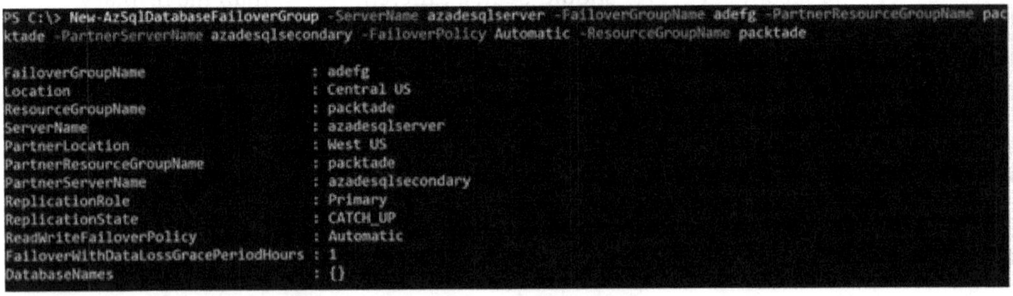

    Figure 2.14 – Creating an auto-failover group

3. Execute the following command to add an existing database in the auto-failover group:

    ```
    $db = Get-AzSqlDatabase -DatabaseName azadesqldb
    -ServerName azadesqlserver -ResourceGroupName packtade
    ```

    ```
    $db | Add-AzSqlDatabaseToFailoverGroup -FailoverGroupName
    adefg
    ```

4. Execute the following command to add a new Azure SQL database to the auto-failover group:

```
$db = New-AzSqlDatabase -DatabaseName azadesqldb2
-Edition basic -ServerName azadesqlserver
-ResourceGroupName packtade
$db | Add-AzSqlDatabaseToFailoverGroup -FailoverGroupName
adefg
```

5. Execute the following PowerShell command to get the details about the auto-failover group:

```
Get-AzSqlDatabaseFailoverGroup -ServerName azadesqlserver
-FailoverGroupName adefg -ResourceGroupName packtade
```

You should get an output as shown in the following screenshot:

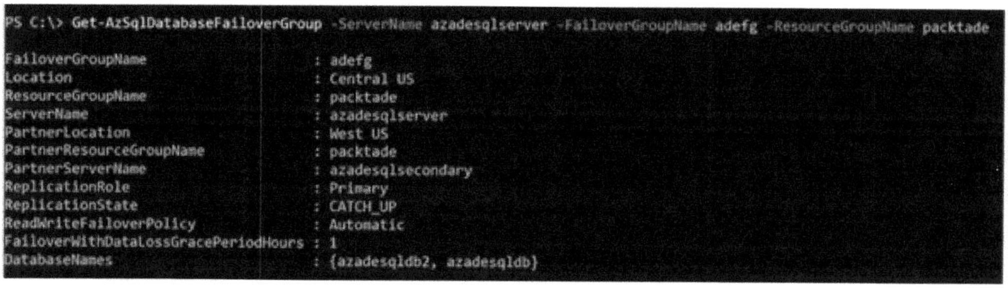

```
PS C:\> Get-AzSqlDatabaseFailoverGroup -ServerName azadesqlserver -FailoverGroupName adefg -ResourceGroupName packtade

FailoverGroupName                     : adefg
Location                              : Central US
ResourceGroupName                     : packtade
ServerName                            : azadesqlserver
PartnerLocation                       : West US
PartnerResourceGroupName              : packtade
PartnerServerName                     : azadesqlsecondary
ReplicationRole                       : Primary
ReplicationState                      : CATCH_UP
ReadWriteFailoverPolicy               : Automatic
FailoverWithDataLossGracePeriodHours  : 1
DatabaseNames                         : {azadesqldb2, azadesqldb}
```

Figure 2.15 – Getting the auto-failover group details

The endpoint used to connect to the primary server of an auto-failover group is of the form `<auto-failover group name>.database.windows.net`. In our case, this will be `adefg.database.windows.net`.

To connect to a readable secondary in an auto-failover group, the endpoint used is of the form `<auto-failover group name>.secondary.database.windows.net`. In our case, the endpoint will be `adefg.secondary.database.windows.net`. In addition to this, we need to specify `ApplicationIntent` as `readonly` in the connection string when connecting to the readable secondary.

6. In an Azure portal, the failover groups can be found on the Azure SQL server page, as shown in the following screenshot:

Figure 2.16 – Viewing an auto-failover group in the Azure portal

7. To open the failover group details, click the failover group name, **adefg**:

Figure 2.17 – Viewing the auto-failover group details in the Azure portal

Performing a manual failover to the secondary server

The steps are as follows:

1. Execute the following command to manually failover to the secondary server:

```
$secondarysqlserver = Get-AzSqlServer -ResourceGroupName
packtade -ServerName azadesqlsecondary
$secondarysqlserver | Switch-AzSqlDatabaseFailoverGroup
-FailoverGroupName adefg
```

If we check in the Azure portal, the primary server is now `azadesqlsecondary` and the secondary server is `azadesqlserver`, as shown in the following screenshot:

Figure 2.18 – Manual failover to the secondary server

2. Execute the following command to remove the auto-failover group. Removing the auto-failover group doesn't remove the secondary or primary SQL databases:

```
Remove-AzSqlDatabaseFailoverGroup -ServerName
azadesqlsecondary -FailoverGroupName adefg
-ResourceGroupName packtade
```

You should get an output as shown in the following screenshot:

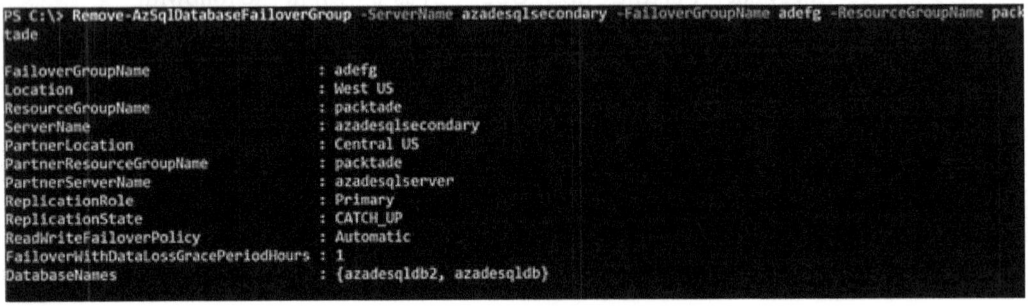

```
PS C:\> Remove-AzSqlDatabaseFailoverGroup -ServerName azadesqlsecondary -FailoverGroupName adefg -ResourceGroupName pack
tade

FailoverGroupName                        : adefg
Location                                 : West US
ResourceGroupName                        : packtade
ServerName                               : azadesqlsecondary
PartnerLocation                          : Central US
PartnerResourceGroupName                 : packtade
PartnerServerName                        : azadesqlserver
ReplicationRole                          : Primary
ReplicationState                         : CATCH_UP
ReadWriteFailoverPolicy                  : Automatic
FailoverWithDataLossGracePeriodHours     : 1
DatabaseNames                            : {azadesqldb2, azadesqldb}
```

Figure 2.19 – Removing the auto-failover group

How it works...

The `New-AzSqlDatabaseFailoverGroup` command is used to create an auto-failover group. We need to specify the auto-failover group name, the primary and secondary server names, the resource group name, and the failover policy (automatic/manual). In addition to this, we can also specify `GracePeriodWithDataLossHours`. As the replication between the primary and secondary is synchronous, the failover may result in data loss. The `GracePeriodwithDataLossHours` value specifies how many hours the system should wait before initiating the automatic failover. This can, therefore, limit the data loss that can happen because of a failover.

After the auto-failover group creation, we can add the databases to the auto-failover group by using the `Add-AzSqlDatabaseToFailoverGroup` command. The database to be added should exist on the primary server and not on the secondary server.

We can perform a manual failover by executing the `Switch-AzSqlDatabaseFailoverGroup` command. We need to provide the primary server name, the auto-failover group name, and the primary server resource group name.

To remove an auto-failover group, execute the `Remove-AzSqlDatabaseFailoverGroup` command by specifying the primary server name and resource group and the auto-failover group name.

Implementing vertical scaling for an Azure SQL database using PowerShell

An Azure SQL database has multiple purchase model and service tiers for different workloads. There are two purchasing models: **DTU-based** and **vCore-based**. There are multiple service tiers within the purchasing models.

Having multiple service tiers gives the flexibility to scale up or scale down based on the workload or activity in an Azure SQL database.

In this recipe, we'll learn how to automatically scale up an Azure SQL database whenever the CPU percentage is above 40%.

Getting ready

In a new PowerShell window, execute the `Connect-AzAccount` command and follow the steps to log in to your Azure account.

You will need an existing Azure SQL database for this recipe. If you don't have one, create an Azure SQL database by following the steps mentioned in the *Provisioning and connecting to an Azure SQL database using PowerShell* recipe.

How to do it...

The steps for this recipe are as follows:

1. Execute the following PowerShell command to create an Azure Automation account:

    ```
    #Create an Azure automation account
    $automation = New-AzAutomationAccount -ResourceGroupName
    packtade -Name adeautomate -Location centralus -Plan
    Basic
    ```

2. Execute the following command to create an Automation runbook of the PowerShell workflow type:

    ```
    #Create a new automation runbook of type PowerShell
    workflow
    $runbook = New-AzAutomationRunbook -Name rnscalesql
    -Description "Scale up sql azure when CPU is 40%"
    -Type PowerShellWorkflow -ResourceGroupName packtade
    -AutomationAccountName $automation.AutomationAccountName
    ```

3. Execute the following command to create Automation credentials. The credentials are passed as a parameter to the runbook and are used to connect to the Azure SQL database from the runbook:

    ```
    #Create automation credentials.
    $sqladminpassword = ConvertTo-SecureString 'Sql@
    Server@1234' -AsPlainText -Force
    ```

```
$sqladmincredential = New-Object System.Management.
Automation.PSCredential ('sqladmin', $sqladminpassword)
```

```
$creds = New-AzAutomationCredential -Name sqlcred
-Description "sql azure creds" -ResourceGroupName
packtade -AutomationAccountName $automation.
AutomationAccountName -Value $sqladmincredential
```

4. The next step is to edit the runbook and PowerShell to modify the service tier of an Azure SQL database. To do that, open `https://portal.azure.com` and log in to your Azure account. Under **All resources**, search for and open the `adeautomate` automation account:

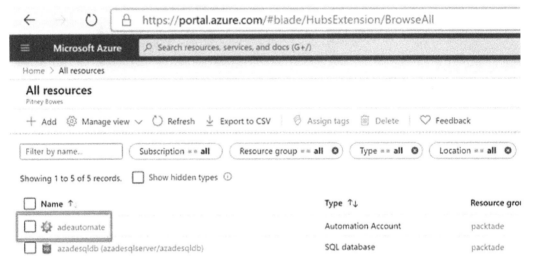

Figure 2.20 – Opening the Azure Automation account

5. On the Azure Automation page, locate and select **Runbooks**:

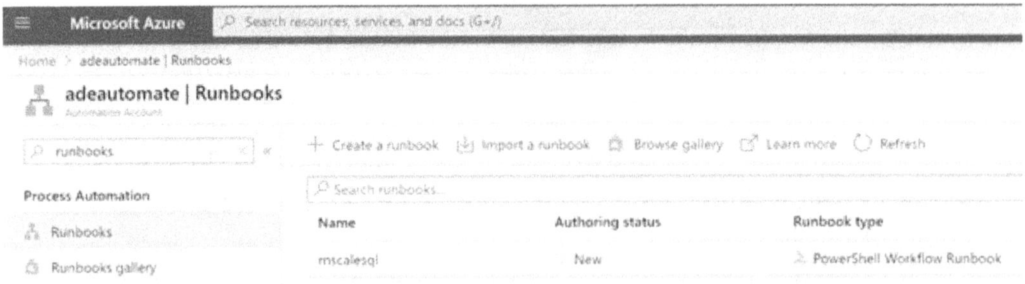

Figure 2.21 – Opening the runbook in Azure Automation

6. Select the `rnscalesql` runbook to open the runbook page. On the runbook page, select **Edit**:

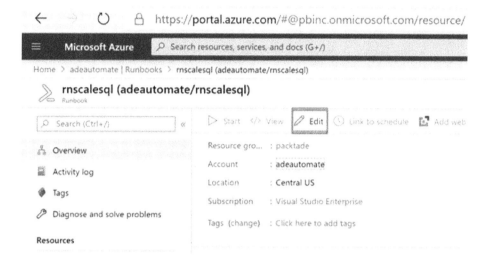

Figure 2.22 – Editing the runbook in your Azure Automation account

7. On the **Edit PowerShell Workflow Runbook** page, copy and paste the following PowerShell code onto the canvas:

```
workflow rnscalesql
{
    param
    (
        # Name of the Azure SQL Database server (Ex:
bzb98er9bp)
        [parameter(Mandatory=$true)]
        [string] $SqlServerName,
        # Target Azure SQL Database name
        [parameter(Mandatory=$true)]
        [string] $DatabaseName,
        # When using in the Azure Automation UI,
please enter the name of the credential asset for the
"Credential" parameter
```

```
        [parameter(Mandatory=$true)]
        [PSCredential] $Credential
    )

    inlinescript
    {
        $ServerName = $Using:SqlServerName + ".database.
windows.net"
        $db = $Using:DatabaseName
        $UserId = $Using:Credential.UserName
        $Password = ($Using:Credential).
GetNetworkCredential().Password
        $ServerName
        $db
        $UserId
        $Password
        $MasterDatabaseConnection = New-Object System.
Data.SqlClient.SqlConnection
        $MasterDatabaseConnection.ConnectionString =
"Server = $ServerName; Database = Master; User ID =
$UserId; Password = $Password;"
        $MasterDatabaseConnection.Open();
        $MasterDatabaseCommand = New-Object System.Data.
SqlClient.SqlCommand
        $MasterDatabaseCommand.Connection =
$MasterDatabaseConnection
        $MasterDatabaseCommand.CommandText =
            "
                ALTER DATABASE $db MODIFY (EDITION =
'Standard', SERVICE_OBJECTIVE = 'S0');

            "
        $MasterDbResult = $MasterDatabaseCommand.
ExecuteNonQuery();
    }
}
```

The preceding code modifies the service tier of the given Azure SQL database to Standard S0.

8. Click **Save**, and then click **Publish** to publish the runbook:

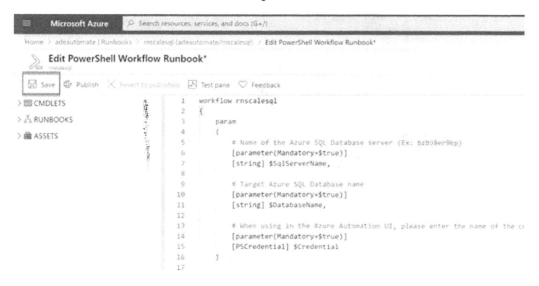

Figure 2.23 – Saving and publishing the runbook

9. The next step is to create a webhook to trigger the runbook. Execute the following command to create the webhook:

```
# define the runbook parameters
$Params = @
{"SQLSERVERNAME"="azadesqlserver";"DATABASENAME"
="azadesqldb";"CREDENTIAL"="sqlcred"}
# Create a webhook
$expiry = (Get-Date).AddDays(1)
New-AzAutomationWebhook -Name whscaleazure -RunbookName
$runbook.Name -Parameters $Params -ResourceGroupName
packtade -AutomationAccountName $automation.
AutomationAccountName -IsEnabled $true -ExpiryTime
$expiry
```

> **Note**
>
> When defining $Params, you may want to change the default values mentioned here if you have a different Azure SQL server, database, and cred values.

You should get an output as shown in the following screenshot:

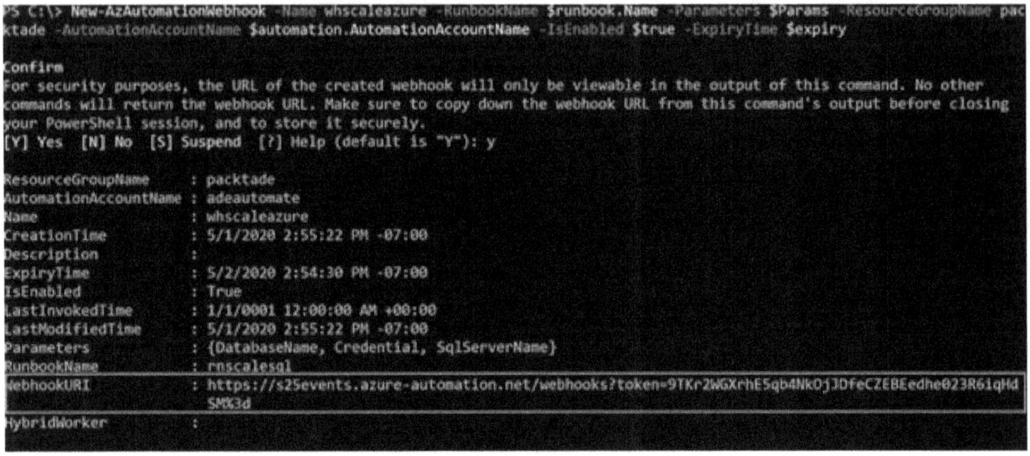

```
PS C:\> New-AzAutomationWebhook -Name whscaleazure -RunbookName $runbook.Name -Parameters $Params -ResourceGroupName pac
ktade -AutomationAccountName $automation.AutomationAccountName -IsEnabled $true -ExpiryTime $expiry

Confirm
For security purposes, the URL of the created webhook will only be viewable in the output of this command. No other
commands will return the webhook URL. Make sure to copy down the webhook URL from this command's output before closing
your PowerShell session, and to store it securely.
[Y] Yes  [N] No  [S] Suspend  [?] Help (default is "Y"): y

ResourceGroupName      : packtade
AutomationAccountName  : adeautomate
Name                   : whscaleazure
CreationTime           : 5/1/2020 2:55:22 PM -07:00
Description            :
ExpiryTime             : 5/2/2020 2:54:30 PM -07:00
IsEnabled              : True
LastInvokedTime        : 1/1/0001 12:00:00 AM +00:00
LastModifiedTime       : 5/1/2020 2:55:22 PM -07:00
Parameters             : {DatabaseName, Credential, SqlServerName}
RunbookName            : rnscalesql
WebhookURI             : https://s25events.azure-automation.net/webhooks?token=9TKr2WGXrhE5qb4NkOjJDfeCZEBEedhe023R61qHd
                         SMk3d
HybridWorker           :
```

Figure 2.24 – Creating a webhook

Copy and save the `WebhookURI` value for later use.

10. The next step is to create an alert for an Azure SQL database that when triggered will call the webhook URI. Execute the following query to create an alert action group receiver:

```
#Create action group reciever
$whr = New-AzActionGroupReceiver -Name agrscalesql
-WebhookReceiver -ServiceUri "https://s25events.azure-
automation.net/webhooks?token=NfL30nj%2fkuSo8TTT7CqDwRI
WEdeXR1lklkK%2fzgELCiY%3d"
```

> **Note**
>
> Replace the value of the `ServiceUri` parameter with your webhook URI from the previous step.

11. Execute the following query to create an action group with an action receiver as defined by the preceding command:

```
#Create a new action group.
$ag = Set-AzActionGroup -ResourceGroupName packtade -Name
ScaleSQLAzure -ShortName scaleazure -Receiver $whr
```

12. Execute the following query to create an alert condition to trigger the alert:

```
#define the alert trigger condition
$condition = New-AzMetricAlertRuleV2Criteria  -MetricName
"cpu_percent" -TimeAggregation maximum -Operator
greaterthan -Threshold 40 -MetricNamespace "Microsoft.
Sql/servers/databases"
```

The condition defines that the alert should trigger when the metric CPU percentage is greater than 40%.

13. Execute the following query to create an alert on the Azure SQL database:

```
#Create the alert with the condition and action defined
in previous steps.
$rid = (Get-AzSqlDatabase -ServerName azadesqlserver
-ResourceGroupName packtade -DatabaseName azadesqldb).
Resourceid
Add-AzMetricAlertRuleV2 -Name monitorcpu
-ResourceGroupName packtade -WindowSize 00:01:00
-Frequency 00:01:00 -TargetResourceId $rid -Condition
$condition  -Severity 1 -ActionGroupId $ag.id
```

You should get an output as shown in the following screenshot:

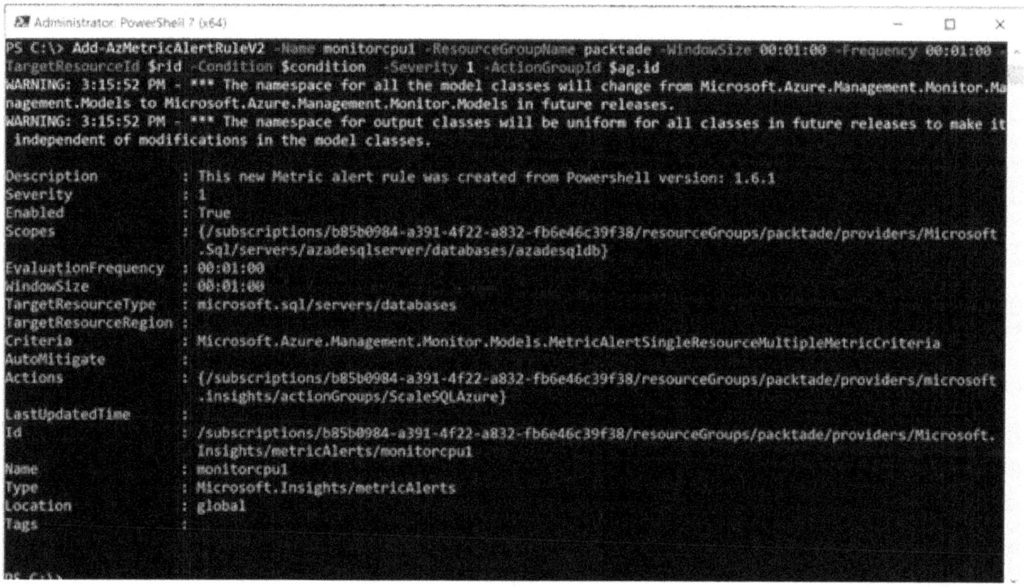

Figure 2.25 – Creating the alert

The preceding command creates an Azure SQL database alert. The alert is triggered when the cpu_percent metric is greater than 40% for more than 1 minute. When the alert is triggered, as defined in the action group, the webhook is called. The webhook in turn runs the runbook. The runbook modifies the service tier of the database to Standard S0.

14. To see the alert in action, connect to the Azure SQL database and execute the following query to simulate high CPU usage:

```
--query to simulate high CPU usage
While(1=1)
Begin
Select cast(a.object_id as nvarchar(max)) from sys.
objects a, sys.objects b,sys.objects c, sys.objects d
End
```

As soon as the alert condition is triggered, the webhook is called and the database service tier is modified to Standard S0.

How it works...

To configure automatic scaling for an Azure SQL database, we create an Azure Automation runbook. The runbook specifies the PowerShell code to modify the service tier of an Azure SQL database.

We create a webhook to trigger the runbook. We create an Azure SQL database alert and define the alert condition to trigger when the cpu_percent metric is greater than 40% for at least 1 minute. In the alert action, we call the webhook defined earlier.

When the alert condition is reached, the webhook is called, which in turn executes the runbook, resulting in the Azure SQL database service tier change.

Implementing an Azure SQL database elastic pool using PowerShell

An elastic pool is a cost-effective mechanism to group single Azure SQL databases of varying peak usage times. For example, consider 20 different SQL databases with varying usage patterns, each Standard S3 requiring 100 **database throughput units** (**DTUs**) to run. We need to pay for 100 DTUs separately. However, we can group all of them in an elastic pool of Standard S3. In this case, we only need to pay for elastic pool pricing and not for each individual SQL database.

In this recipe, we'll create an elastic pool of multiple single Azure databases.

Getting ready

In a new PowerShell window, execute the `Connect-AzAccount` command and follow the steps to log in to your Azure account.

How it works...

The steps for this recipe are as follows:

1. Execute the following query on an Azure SQL server:

    ```
    #create credential object for the Azure SQL Server admin
    credential

    $sqladminpassword = ConvertTo-SecureString 'Sql@
    Server@1234' -AsPlainText -Force

    $sqladmincredential = New-Object System.Management.
    Automation.PSCredential ('sqladmin', $sqladminpassword)

    # create the azure sql server

    New-AzSqlServer -ServerName azadesqlserver
    -SqlAdministratorCredentials $sqladmincredential
    -Location "central us" -ResourceGroupName packtade

    Execute the following query to create an elastic pool.

    #Create an elastic pool

    New-AzSqlElasticPool -ElasticPoolName adepool -ServerName
    azadesqlserver -Edition standard -Dtu 100 -DatabaseDtuMin
    20 -DatabaseDtuMax 100 -ResourceGroupName packtade
    ```

    You should get an output as shown in the following screenshot:

Figure 2.26 – Creating a new Azure elastic pool

2. Execute the following query to create and add an Azure SQL database to an elastic pool:

```
#Create a new database in elastic pool
New-AzSqlDatabase -DatabaseName azadedb1 -ElasticPoolName
adepool -ServerName azadesqlserver -ResourceGroupName
packtade
```

You should get an output as shown in the following screenshot:

Figure 2.27 – Creating a new SQL database in an elastic pool

3. Execute the following query to create a new Azure SQL database outside of the elastic pool:

```
#Create a new database outside of an elastic pool
New-AzSqlDatabase -DatabaseName azadedb2 -Edition
Standard -RequestedServiceObjectiveName S3 -ServerName
azadesqlserver -ResourceGroupName packtade
```

You should get an output as shown in the following screenshot:

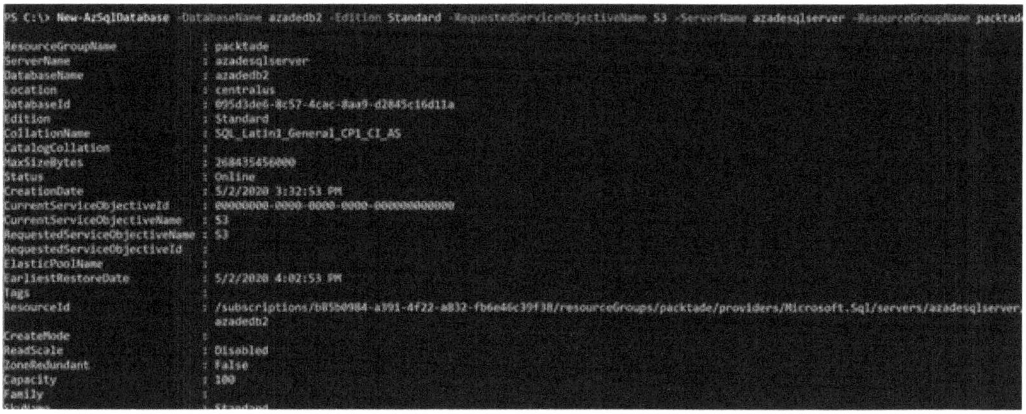

Figure 2.28 – Creating a new SQL database

4. Execute the following query to add the `adesqldb2` database created in the preceding step to the elastic pool:

```
#Add an existing database to the elastic pool
$db = Get-AzSqlDatabase -DatabaseName azadedb2
-ServerName azadesqlserver -ResourceGroupName packtade
$db | Set-AzSqlDatabase -ElasticPoolName adepool
```

You should get an output as shown in the following screenshot:

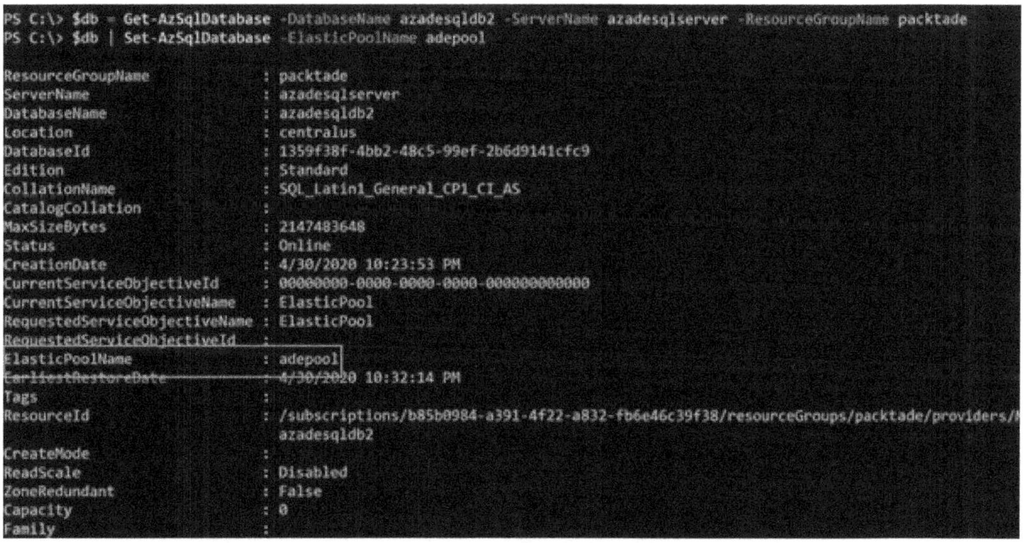

Figure 2.29 – Adding an existing SQL database to an elastic pool

5. To verify this in the Azure portal, log in with your Azure account. Navigate to **All resources | azadesqlserver | SQL elastic pools | Configure**:

Figure 2.30 – Viewing the elastic pool in the Azure portal

6. Execute the following command to remove an Azure SQL database from an elastic pool. To move a database out of an elastic pool, we need to set the edition and the service objective explicitly:

```
#remove a database from an elastic pool
$db = Get-AzSqlDatabase -DatabaseName azadesqldb2
-ServerName azadesqlserver -ResourceGroupName packtade

$db | Set-AzSqlDatabase -Edition Standard
-RequestedServiceObjectiveName S3
```

You should get an output as shown in the following screenshot:

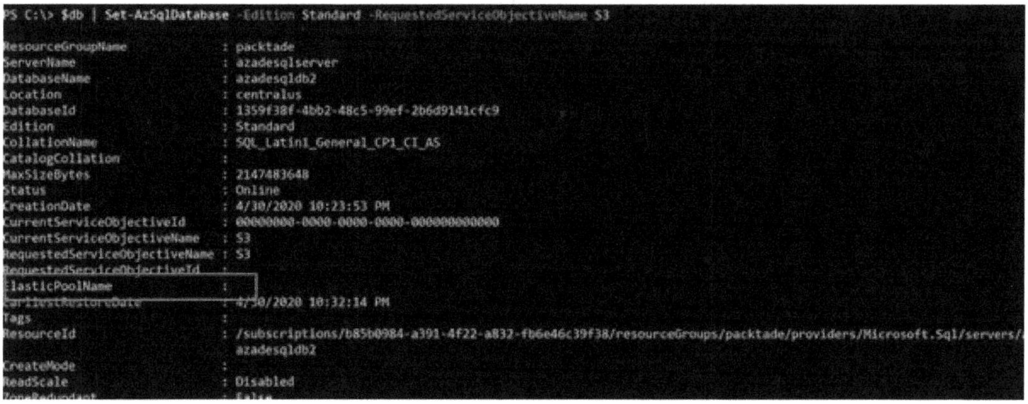

Figure 2.31 – Removing a SQL database from an elastic pool

7. Execute the command that follows to remove an elastic pool. An elastic pool should be empty before it can be removed. Execute the following query to remove all of the databases in an elastic pool:

```
# get elastic pool object
$epool = Get-AzSqlElasticPool -ElasticPoolName adepool
-ServerName azadesqlserver -ResourceGroupName packtade
# get all databases in an elastic pool
$epdbs = $epool | Get-AzSqlElasticPoolDatabase
# change the edition of all databases in an elastic pool
to standard S3
foreach($db in $epdbs) {
$db | Set-AzSqlDatabase -Edition Standard
-RequestedServiceObjectiveName S3
}
# Remove an elastic pool
$epool | Remove-AzSqlElasticPool
```

> **Note**
>
> The command sets the edition of the SQL databases to Standard. This is for demo purposes only. If this is to be done on production, modify the edition and service objective accordingly.

How it works...

We create an elastic pool using the `New-AzSqlElasticPool` command. In addition to the parameters, such as the server name, resource group name, compute model, compute generation, and edition, which are the same as when we create a new Azure SQL database, we can specify `DatabaseMinDtu` and `DatabaseMaxDtu`. `DatabaseMinDtu` specifies the minimum DTU that all databases can have in an elastic pool. `DatabaseMaxDtu` is the maximum DTU that a database can consume in an elastic pool.

Similarly, for the vCore-based purchasing model, we can specify `DatabaseVCoreMin` and `DatabaseVCoreMax`.

To add a new database to an elastic pool, specify the elastic pool name at the time of database creation using the `New-AzSqlDatabase` command.

To add an existing database to an elastic pool, modify the database using `Set-AzSqlDatabase` to specify the elastic pool name.

To remove a database from an elastic pool, modify the database using the `Set-AzSqlDatabase` command to specify a database edition explicitly.

To remove an elastic pool, first empty it by moving out all of the databases from the elastic pool, and then remove it using the `Remove-AzSqlElasticPool` command.

Monitoring an Azure SQL database using the Azure portal

Azure SQL Database has built-in monitoring features, such as query performance insights, performance overview, and diagnostic logging. In this recipe, we'll learn how to use the monitoring capabilities using the Azure portal.

Getting ready

We'll use PowerShell to create an Azure SQL database, so open a PowerShell window and log in to your Azure account by executing the `Connect-AzAccount` command.

We'll use the Azure portal to monitor the Azure SQL database. Open `https://portal.azure.com` and log in to your Azure account.

How to do it...

First, let's execute a sample workload.

Creating an Azure SQL database and executing a sample workload

The steps are as follows:

1. Execute the following PowerShell command to create an Azure SQL database with the `AdventureWorksLT` sample database:

```
# create the resource group
New-AzResourceGroup -Name packtade -Location "central us"
-force
#create credential object for the Azure SQL Server admin
credential
$sqladminpassword = ConvertTo-SecureString 'Sql@
Server@1234' -AsPlainText -Force
$sqladmincredential = New-Object System.Management.
Automation.PSCredential ('sqladmin', $sqladminpassword)
# create the azure sql server
New-AzSqlServer -ServerName azadesqlserver
-SqlAdministratorCredentials $sqladmincredential
-Location "central us" -ResourceGroupName packtade
#Create the SQL Database
New-AzSqlDatabase -DatabaseName adeawlt -Edition basic
-ServerName azadesqlserver -ResourceGroupName packtade
-SampleName AdventureWorksLT
```

2. Execute the following command to add the client IP to the Azure SQL Server firewall:

```
$clientip = (Invoke-RestMethod -Uri https://ipinfo.io/
json).ip
New-AzSqlServerFirewallRule -FirewallRuleName "home"
-StartIpAddress $clientip -EndIpAddress $clientip
-ServerName azadesqlserver -ResourceGroupName packtade
```

3. Execute the following command to run a workload against the Azure SQL database:

```
sqlcmd -S azadesqlserver.database.windows.net -d adeawlt
-U sqladmin -P Sql@Server@1234 -i "C:\ADECookbook\
Chapter02\workload.sql" > "C:\ADECookbook\Chapter02\
workload_output.txt"
```

It can take 4–5 minutes for the workload to complete. You can execute the preceding command multiple times; however, you should run it at least once.

Monitoring Azure SQL database metrics

The steps are as follows:

1. In the Azure portal, navigate to **All resources** | **azadesqlserver** | the **adeawlt** database. Search for and open **Metrics**:

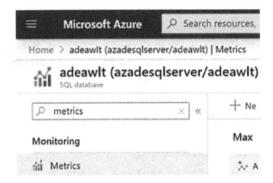

Figure 2.32 – Opening the Metrics section in the Azure portal

The **Metrics** page allows you to monitor different available metrics over time.

2. To select metrics, click **Add metric** | **CPU percentage** | **Data IO percentage**:

Figure 2.33 – Monitoring metrics for a SQL database

We can select the metrics we are interested in monitoring and use the **Pin to dashboard** feature to pin the chart to the portal dashboard. We can also create an alert rule from the metrics page by clicking on **New alert rule**. We can select a time range to drill down to specific times or investigates spikes in the chart.

3. To select a time range, select the **Time range** dropdown in the top-right corner of the **Metrics** page and select the desired time range:

Figure 2.34 – Selecting a time range to monitor

Using Query Performance Insight to find resource-consuming queries

Query Performance Insight is an intelligent performance feature that allows us to find any resource-consuming and long-running queries. The steps are as follows:

1. In the Azure portal, navigate to **All resources | azadesqlserver** | the **adeawlt** database. Find and open **Query Performance Insight**:

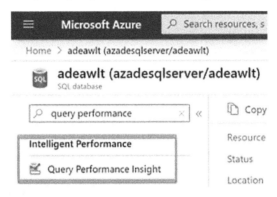

Figure 2.35 – Selecting Query Performance Insight for the SQL database

2. On the **Query Performance Insight** page, observe that there are three tabs: **Resource Consuming Queries**, **Long Running Queries**, and **Custom**. We can select resource-consuming queries by **CPU**, **Data IO**, and **Log IO**:

Figure 2.36 – Monitoring queries for the SQL database

Resource Consuming Queries lists out the top three queries by CPU consumption. We can also select the top three queries by **Data IO** and **Log IO**. The bottom of the page lists out the color-coded queries.

3. To get the query text, click on the color-coded box:

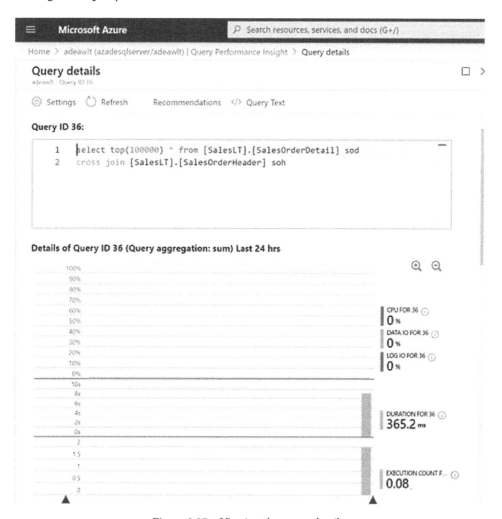

Figure 2.37 – Viewing the query details

We can look at the query text and optimize it for better performance.

4. The **Custom** tab allows us to select resource-consuming queries by duration and execution count. We can also specify a custom time range, the number of queries, and the query and metric aggregation:

Figure 2.38 – Providing custom monitoring configuration

5. Select the options and click the **Go** button to refresh the chart as per the custom settings. **Long running queries** lists out the top three queries by duration:

Figure 2.39 – Viewing the long-running queries list

We can further look into the query text and other details by selecting the query ID.

Monitoring an Azure SQL database using diagnostic settings

In addition to metrics and query performance insight, we can also monitor an Azure SQL database by collecting diagnostic logs. The diagnostic logs can be sent to the Log Analytics workspace or Azure Storage, or can be streamed to Azure Event Hubs. The steps are as follows:

1. To enable diagnostic logging using the Azure portal, navigate to **All resources |
    azadesqlserver | adeawlt**. Find and open **Diagnostic settings**:

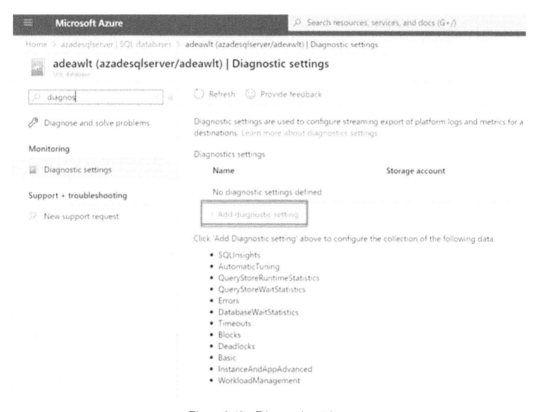

Figure 2.40 – Diagnostic settings

2. Click on **Add diagnostic setting** to add a new diagnostic setting.

3. Select the categories to be included in the logs and their destination:

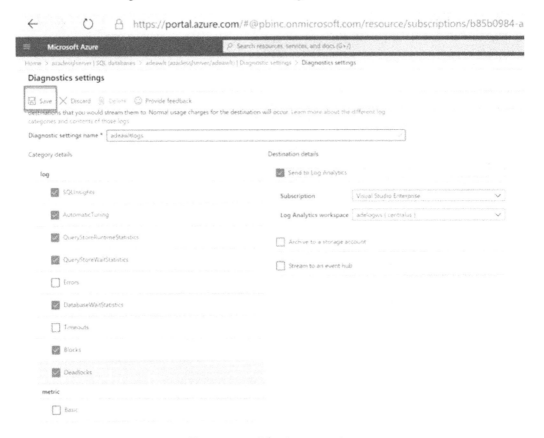

Figure 2.41 – Selecting categories

4. Click **Save** to create the new diagnostic setting. The diagnostic logs can be analyzed in the Log Analytics workspace.

> **Note:**
>
> **Diagnostic setting** adds an additional cost to the Azure SQL database. It may take some time for the logs to be available after creating a new diagnostic setting.

Automatic tuning in an Azure SQL database

Automatic tuning provides three features: **force plan**, **create**, and **drop indexes**. Automatic tuning can be enabled for an Azure SQL server, in which case it's applied to all of the databases in that Azure SQL server. Automatic tuning can be enabled for individual Azure SQL databases as well. The steps are as follows:

1. To enable automatic tuning, in the Azure portal, navigate to **All resources | azadesqlserver | adewlt**. Find and select **Automatic Tuning** under **Intelligent Performance**:

Figure 2.42 – Automatic tuning in the SQL database

2. Enable the **CREATE INDEX** tuning option by clicking **ON** under the **Desired state** option.

3. Click **Apply** to save the configuration.

> **Note**
> It may take time for recommendations to show up.

The recommendations will show up in the performance recommendations under the **Intelligent Performance** section.

3
Analyzing Data with Azure Synapse Analytics

Azure Synapse Analytics, formerly known as **SQL Data Warehouse**, combines data warehousing and big data analytics to provide a unified experience to extract, load, and transform data. Azure Synapse has the following features—Synapse SQL – T-SQL-based analytics (SQL pools and SQL on-demand), Spark Analytics, Synapse pipelines, and Synapse studio. At the time of writing this book, all features except Synapse SQL are in preview.

Azure Synapse Analytics is important to learn as data warehousing is an important part of data engineering and big data solutions.

Azure Synapse SQL can be used to quickly load data from sources and perform transformations. The transformation queries are fast as Azure Synapse SQL uses **massive parallel processing** (**MPP**) architecture to process the queries.

In this chapter, we'll cover the following recipes:

- Provisioning and connecting to an Azure Synapse SQL pool using PowerShell
- Pausing or resuming a Synapse SQL pool using PowerShell
- Scaling an Azure Synapse SQL pool instance using PowerShell
- Loading data into a SQL pool using PolyBase with T-SQL
- Loading data into a SQL pool using the COPY INTO statement
- Implementing workload management in an Azure Synapse SQL pool
- Optimizing queries using materialized views in Azure Synapse Analytics

Technical requirements

The following tools are required for the recipes in this chapter:

- A Microsoft Azure subscription
- PowerShell 7
- Microsoft Azure PowerShell
- **SQL Server Management Studio** (**SSMS**) or Azure Data Studio

Provisioning and connecting to an Azure Synapse SQL pool using PowerShell

In this recipe, we'll provision an Azure Synapse SQL pool using PowerShell. Provisioning a Synapse SQL pool uses the same commands that are used for provisioning an Azure SQL database, but with different parameters.

Getting ready

Before you start, log in to Azure from PowerShell. To do this, execute the following command and follow the instructions to log in to Azure:

```
Connect-AzAccount
```

How to do it...

Follow the given steps to provision a new Azure Synapse SQL pool:

1. Execute the following command to create a new resource group:

    ```
    #Create resource group
    New-AzResourceGroup -Name packtade -location centralus
    ```

2. Execute the following command to create a new Azure SQL server:

    ```
    #create credential object for the Azure SQL Server admin
    credential
    $sqladminpassword = ConvertTo-SecureString 'Sql@
    Server@1234' -AsPlainText -Force
    $sqladmincredential = New-Object System.Management.
    Automation.PSCredential ('sqladmin', $sqladminpassword)
    # create the azure sql server
    New-AzSqlServer -ServerName azadesqlserver
    -SqlAdministratorCredentials $sqladmincredential
    -Location "central us" -ResourceGroupName packtade
    ```

    You should get an output as shown in the following screenshot:

    Figure 3.1 – Creating a new Azure SQL server

3. Execute the following command to create a new Azure Synapse SQL pool:

    ```
    New-AzSqlDatabase -ServerName azadesqlserver
    -DatabaseName azadesqldw -Edition DataWarehouse
    -RequestedServiceObjectiveName DW100c -ResourceGroupName
    packtade
    ```

You should get an output as shown in the following screenshot:

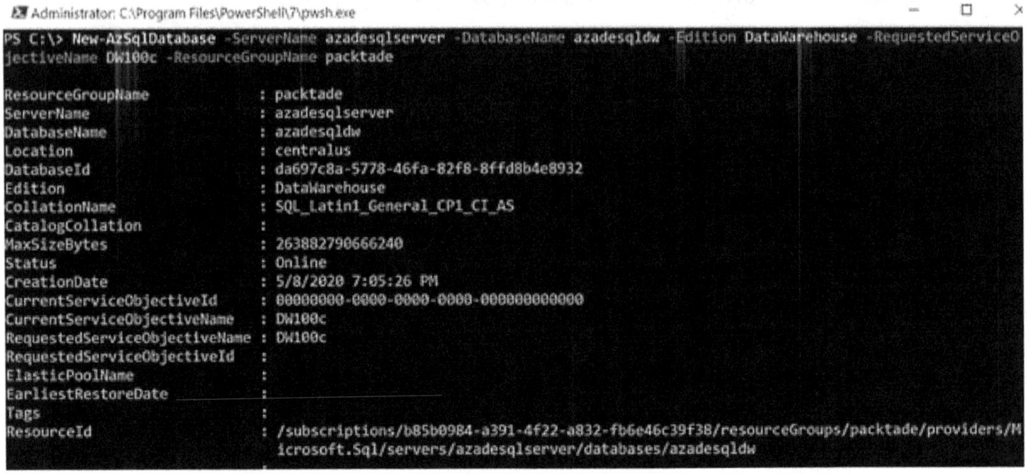

Figure 3.2 – Creating a new Azure Synapse SQL pool

4. Execute the following command to whitelist the public IP of the client machine in the Azure SQL Server firewall:

```
#Get the client public ip
$clientip = (Invoke-RestMethod -Uri https://ipinfo.io/
json).ip
#Whitelist the public IP in firewall
New-AzSqlServerFirewallRule -FirewallRuleName "home"
-StartIpAddress $clientip -EndIpAddress $clientip
-ServerName azadesqlserver -ResourceGroupName packtade
```

You should get the following output:

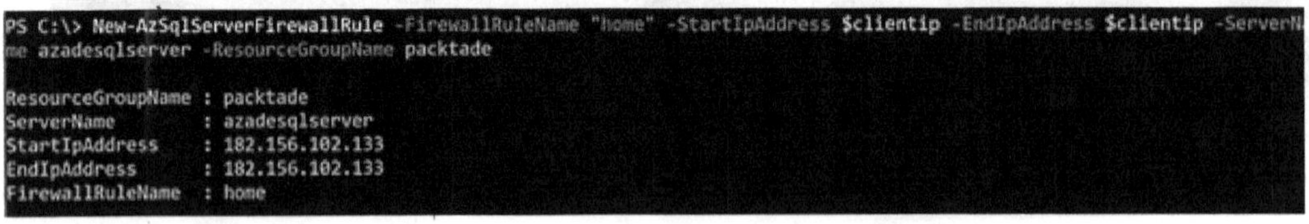

Figure 3.3 – Adding a client public IP in the Azure SQL Server firewall

5. Execute the following command to connect to the Synapse SQL pool:

```
sqlcmd -S "azadesqlserver.database.windows.net" -U
sqladmin -P "Sql@Server@1234" -d azadesqldw -I
```

When connecting to a Synapse SQL pool, the **Quoted Identifier** session setting should be set to **ON**. The -I switch in the sqlcmd command switches on the quoted identifier session setting.

6. Execute the following command to remove Azure SQL Server and the Azure Synapse SQL pool:

```
Remove-AzSqlServer -ServerName azadesqlserver
-ResourceGroupName packtade
```

You should get an output as shown in the following screenshot:

Figure 3.4 – Removing an Azure SQL server

How it works...

To provision an Azure Synapse SQL pool, we first need to create the logical Azure SQL server. To create an Azure SQL server, we use the New-AzSqlServer command. We provide the server name, resource group, location, and admin credentials as parameters.

To create an Azure Synapse SQL pool, we use the New-AzSqlDatabase command. This is the same command that is used to provision an Azure SQL database. We provide the server name, database name, resource group name, edition, and service objective. In the **Edition** parameter, we provide **Data warehouse**.

To add a new firewall rule, we use the New-AzSqlServerFirewallRule command. We provide a firewall rule name, server name, resource group name, and the start and end IP as parameters. Removing Azure SQL Server also removes the Azure Synapse SQL pool.

To connect to an Azure Synapse SQL pool, we can use sqlcmd, SSMS, Azure Data Studio, or any of the valid SQL drivers. The server name to be provided should be of the form <servername>.database.windows.net.

To remove an Azure SQL server, we use the Remove-AzSqlServer command. We provide the server name and resource group name as parameters.

Pausing or resuming a Synapse SQL pool using PowerShell

In an Azure Synapse SQL pool, the compute and storage are decoupled and are therefore charged separately. This means that if we aren't running any queries and are not using the compute, we can pause the Synapse SQL pool to save on compute cost. We'll still be charged for the storage.

A data warehouse is mostly used when it's to be refreshed with the new data or any computation needs to be performed to prepare the summary tables for the reports. We can save compute costs when it's not being used.

In this recipe, we'll learn how to pause and resume a Synapse SQL pool.

Getting ready

Before you start, log in to Azure from PowerShell. To do this, execute the following command and follow the instructions to log in to Azure:

```
Connect-AzAccount
```

You need a Synapse SQL pool to perform the steps in this recipe. If you don't have an existing Synapse SQL pool, you can create one using the steps from the *Provisioning and connecting to an Azure Synapse SQL pool using PowerShell* recipe.

How to do it...

The steps for this recipe are as follows:

1. Execute the following command to check the status of the Synapse SQL pool:

    ```
    #get the Synapse SQL pool object
    $dw = Get-AzSqlDatabase -ServerName azadesqlserver
    -DatabaseName azadesqldw -ResourceGroupName packtade
    #check the status
    $dw.Status
    ```

    You should get an output as shown in the following screenshot:

    ```
    PS C:\> $dw = Get-AzSqlDatabase -ServerName azadesqlserver -DatabaseName azadesqldw -ResourceGroupName packtade
    PS C:\> $dw.Status
    Online
    ```

    Figure 3.5 – Getting the Azure Synapse SQL pool status

2. Execute the following command to suspend the Synapse SQL pool:

```
#suspend the Synapse SQL pool compute
$dw | Suspend-AzSqlDatabase
```

You should get an output as shown in the following screenshot:

```
PS C:\> $dw | Suspend-AzSqlDatabase

ResourceGroupName            : packtade
ServerName                   : azadesqlserver
DatabaseName                 : azadesqldw
Location                     : Central US
DatabaseId                   : 7afa8923-ff36-4296-b520-86a12db946d8
Edition                      : DataWarehouse
CollationName                : SQL_Latin1_General_CP1_CI_AS
CatalogCollation             :
MaxSizeBytes                 : 263882790666240
Status                       : Paused
CreationDate                 : 5/8/2020 8:50:41 PM
CurrentServiceObjectiveId    : 9f848803-41b2-4a6d-9501-bb0e0ab31d2e
CurrentServiceObjectiveName  : DW100c
RequestedServiceObjectiveName :
```

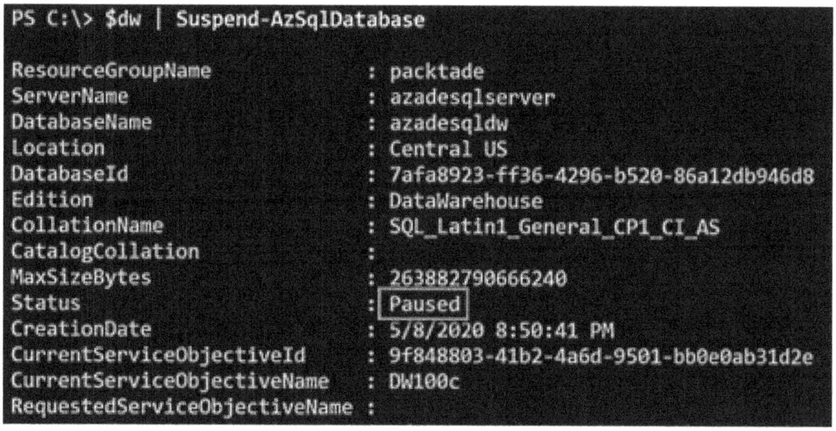

Figure 3.6 – Pausing an Azure Synapse SQL pool

3. Execute the following command to resume the Synapse SQL pool:

```
$dw | Resume-AzSqlDatabase
```

You should get an output as shown in the following screenshot:

```
PS C:\> $dw | Resume-AzSqlDatabase

ResourceGroupName            : packtade
ServerName                   : azadesqlserver
DatabaseName                 : azadesqldw
Location                     : Central US
DatabaseId                   : 7afa8923-ff36-4296-b520-86a12db946d8
Edition                      : DataWarehouse
CollationName                : SQL_Latin1_General_CP1_CI_AS
CatalogCollation             :
MaxSizeBytes                 : 263882790666240
Status                       : Online
CreationDate                 : 5/8/2020 8:50:41 PM
CurrentServiceObjectiveId    : 9f848803-41b2-4a6d-9501-bb0e0ab31d2e
CurrentServiceObjectiveName  : DW100c
RequestedServiceObjectiveName :
RequestedServiceObjectiveId  : 9f848803-41b2-4a6d-9501-bb0e0ab31d2e
ElasticPoolName              :
```

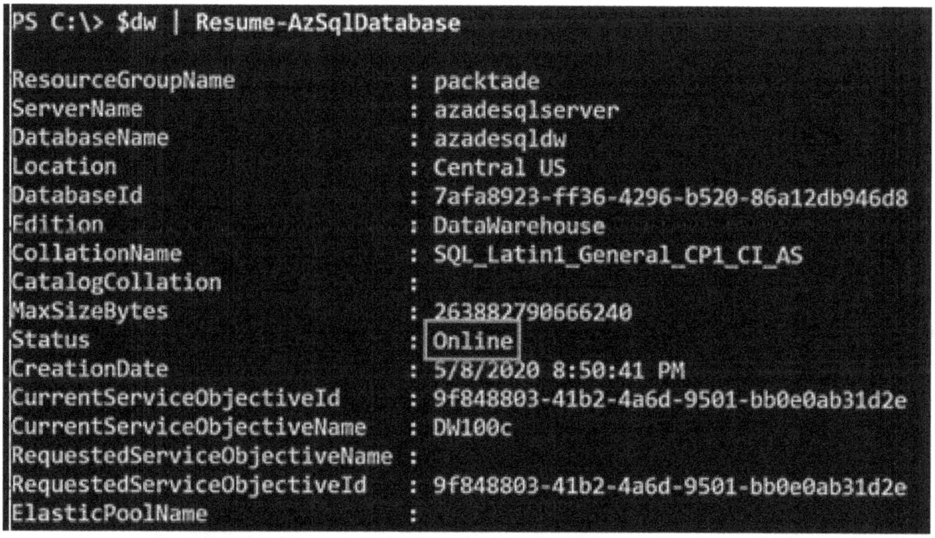

Figure 3.7 – Resuming an Azure Synapse SQL pool

How it works...

To suspend an Azure Synapse SQL pool, use `Suspend-AzSqlDatabase`. To resume an Azure Synapse SQL pool, use `Resume-AzSqlDatabase`. The two commands accept the same parameters: server name, database name, and resource group name. We can also pipe the database (SQL pool) object to the command.

To check the status of the Synapse SQL pool, check the `Status` property from the `Get-AzSqlDatabase` command.

This can help with saving costs. For example, if the data warehouse isn't used at night or on weekends, or it's only used at specified times, we can use Azure Automation or any other scheduling tool to suspend or resume the Synapse SQL pool based on the usage schedule.

Scaling an Azure Synapse SQL pool instance using PowerShell

An Azure Synapse SQL pool can be scaled up or down as per the usage or the workload requirement. For example, consider a scenario where we are at performance level DW100c. A new workload requires a higher performance level. Therefore, to support the new workload, we can scale up to a higher performance level, say, DW400c, and scale down to DW100c when the workload finishes.

In this recipe, we'll learn how to scale up or scale down an Azure Synapse SQL pool using PowerShell.

Getting ready

Before you start, log in to Azure from PowerShell. To do this, execute the following command and follow the instructions to log in to Azure:

```
Connect-AzAccount
```

You need a Synapse SQL pool to perform the steps in the recipe. If you don't have an existing Synapse SQL pool, you can create one using the steps from the *Provisioning and connecting to an Azure Synapse SQL pool using PowerShell* recipe.

How to do it...

The steps for this recipe are as follows:

1. Execute the following command to get the performance tier of an Azure Synapse SQL pool:

```
$dw = Get-AzSqlDatabase -ServerName azadesqlserver
-DatabaseName azadesqldw -ResourceGroupName packtade
$dw.CurrentServiceObjectiveName
```

You should get an output as follows:

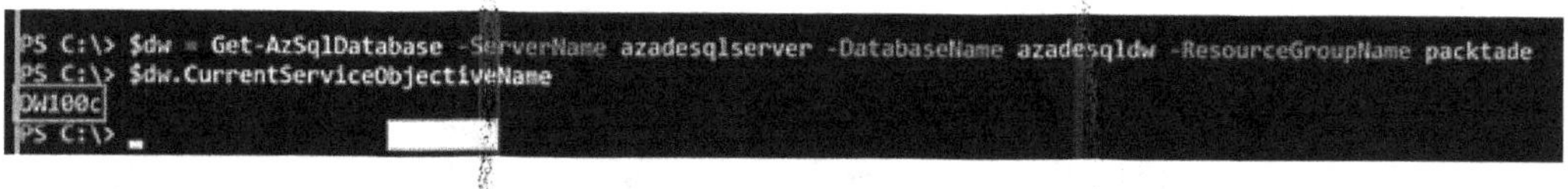

Figure 3.8 – Getting the performance tier of a Synapse SQL pool

2. Execute the following command to scale up the performance tier to DW200c from DW100c:

```
$dw | Set-AzSqlDatabase -RequestedServiceObjectiveName
DW200c
```

You should get an output as shown in the following screenshot:

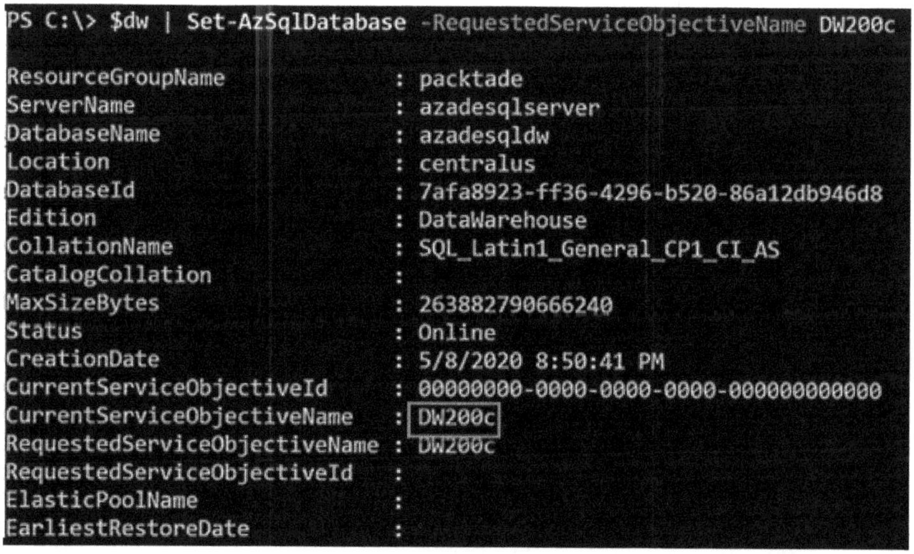

Figure 3.9 – Scaling up a Synapse SQL pool

3. Execute the following command to scale down the Synapse SQL pool to DW100c:

```
$dw | Set-AzSqlDatabase -RequestedServiceObjectiveName
DW100c
```

You should get an output as shown in the following screenshot:

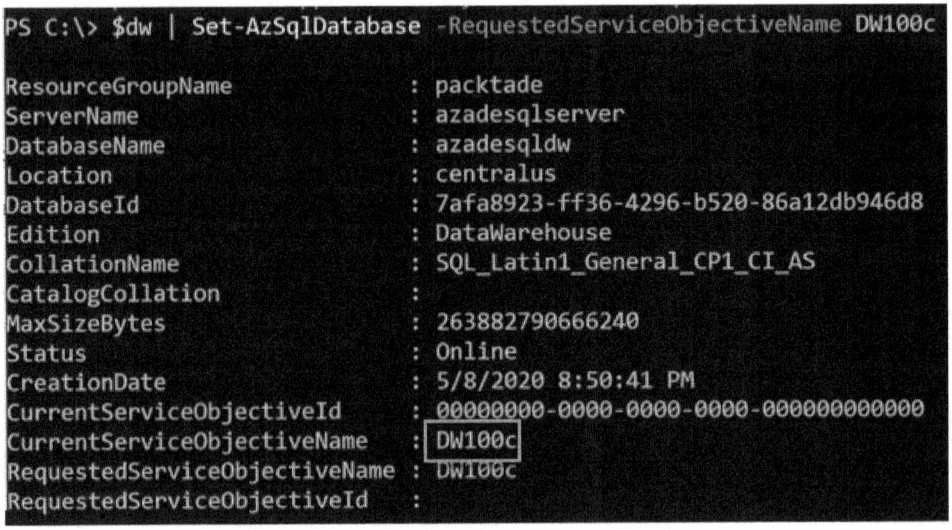

```
PS C:\> $dw | Set-AzSqlDatabase -RequestedServiceObjectiveName DW100c

ResourceGroupName             : packtade
ServerName                    : azadesqlserver
DatabaseName                  : azadesqldw
Location                      : centralus
DatabaseId                    : 7afa8923-ff36-4296-b520-86a12db946d8
Edition                       : DataWarehouse
CollationName                 : SQL_Latin1_General_CP1_CI_AS
CatalogCollation              :
MaxSizeBytes                  : 263882790666240
Status                        : Online
CreationDate                  : 5/8/2020 8:50:41 PM
CurrentServiceObjectiveId     : 00000000-0000-0000-0000-000000000000
CurrentServiceObjectiveName   : DW100c
RequestedServiceObjectiveName : DW100c
RequestedServiceObjectiveId   :
```

Figure 3.10 – Scaling down a Synapse SQL pool

How it works...

To scale up or down an Azure Synapse SQL pool, use the Set-AzSqlDatabase command. The RequestedServiceObjectiveName parameter defines the performance tier. The Set-AzSqlDatabase command can also be used to modify other properties or configuration settings of an Azure Synapse SQL pool.

Loading data into a SQL pool using PolyBase with T-SQL

PolyBase allows you to query external data in Hadoop, Azure Blob storage, or Azure Data Lake Storage from SQL Server using T-SQL. In this recipe, we'll import data from a CSV file in an Azure Data Lake Storage account into an Azure Synapse SQL pool using PolyBase.

Getting ready

Before you start, log in to Azure from PowerShell. To do this, execute the following command and follow the instructions to log in to Azure:

```
Connect-AzAccount
```

You need a Synapse SQL pool to perform the steps in the recipe. If you don't have an existing Synapse SQL pool, you can create one using the steps from the *Provisioning and connecting to an Azure Synapse SQL pool using PowerShell* recipe.

How to do it...

Follow the given steps to import data into Azure Synapse SQL using PolyBase:

1. Execute the following command to create an Azure Data Lake Storage account and upload the data:

```
#Create a new Azure Data Lake Storage account
$dlsa = New-AzStorageAccount -ResourceGroupName
packtade -Name azadedls -SkuName Standard_LRS
-Location centralus -Kind StorageV2 -AccessTier Hot
-EnableHierarchicalNamespace $true
#create the file system (container)
New-AzDatalakeGen2FileSystem -Name ecommerce -Context
$dlsa.Context
#Create the directory within the file system
New-AzDataLakeGen2Item -FileSystem ecommerce -Path
"orders/" -Context $dlsa.Context -Directory
#upload files to the directory
#get all files to be uploaded
$files = Get-ChildItem -Path C:\ADECookbook\Chapter03\
Data\
#upload all files in the data directory to Azure data
lake storage
  foreach($file in $files) {
$source = $file.FullName
$dest = "orders/" + $file.Name
New-AzDataLakeGen2Item -FileSystem ecommerce -Path
$dest -Source $source -Context $dlsa.Context
-ConcurrentTaskCount 50 -Force
}
```

```
#list out all the files from the orders directory in
Azure data lake storage
Get-AzStorageBlob -Container ecommerce -Context $dlsa.
Context
```

You should get an output as shown in the following screenshot:

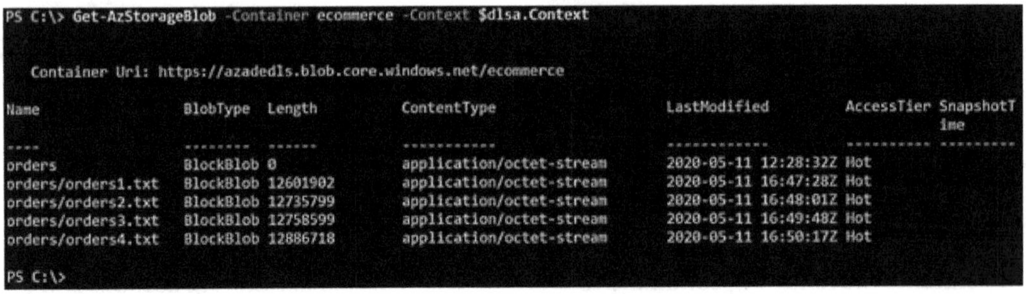

Figure 3.11 – Listing all blobs in an Azure Data Lake Storage container

2. We need an Azure Data Lake Storage account key for the later steps. Execute the following command to get the key value and save it for later steps:

```
$dlsa | Get-AzStorageAccountKey
```

You should get an output as shown in the following screenshot:

Figure 3.12 – Getting the account key for an Azure Data Lake Storage account

Copy and save the value of key1 to be used later.

3. Execute the following command to whitelist the client IP in the Azure SQL Server firewall:

```
$clientip = (Invoke-RestMethod -Uri https://ipinfo.io/
json).ip
New-AzSqlServerFirewallRule -FirewallRuleName "home"
-StartIpAddress $clientip -EndIpAddress $clientip
-ServerName azadesqlserver -ResourceGroupName packtade
```

4. Open SSMS and connect to the Azure Synapse SQL pool in Object Explorer:

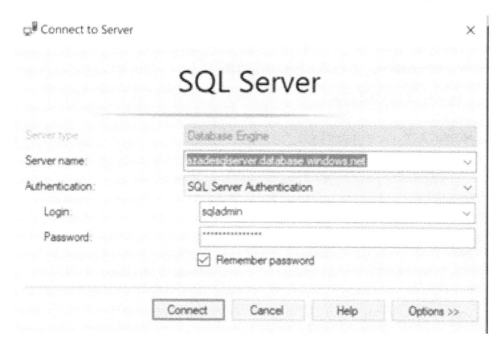

Figure 3.13 – Connecting to the Synapse SQL pool from SSMS

5. From the **Object Explorer** panel, expand **Databases** and right-click the **azadesqldw** SQL pool. From the right-click menu, select **New Query**:

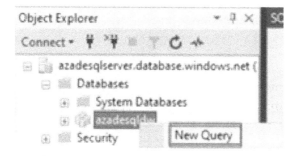

Figure 3.14 – Opening a new query window in SSMS

In the new query window, execute the steps that follow to load the files uploaded in *step 1* into a table in the Azure Synapse SQL pool.

6. Execute the following T-SQL to create a database master key:

```
CREATE MASTER KEY;
GO
```

7. Execute the following command to create database-scoped credentials:

```
CREATE DATABASE SCOPED CREDENTIAL ADLSCreds
WITH
IDENTITY = 'azadedls',
SECRET = 'RmM0cmNi0SUZ1OP3yy6HW61xg9KvjWW1Mui9ET/
1f41duSi8SsVItYtwiwvH4R7aku0DGY/KFG6qW/p5gHocvQ=='
```

The value of the SECRET parameter is the Azure Data Lake Storage account key value we saved in *step 2*.

> **Note**
> The Azure Data Lake Storage account key will be different in your case.

8. Execute the following command to create the external data source:

```
CREATE EXTERNAL DATA SOURCE adls
WITH (
TYPE = HADOOP,
LOCATION='abfs://ecommerce@azadedls.dfs.core.windows.
net',
CREDENTIAL = ADLSCreds
);
```

The external data source is the Azure Data Lake Storage account created in *step 1*. The location parameter is of the following format: abfs://containername@azuredatalakestorageaccountname.dfs.core.windows.net

> **Note**
> Use abfss instead of abfs if the Azure account has secure transfer enabled.

9. Execute the following query to create the external file format (drop the orderfileformat external file format):

```
CREATE EXTERNAL FILE FORMAT orderfileformat
WITH
```

```
( FORMAT_TYPE = DELIMITEDTEXT
, FORMAT_OPTIONS ( FIELD_TERMINATOR = '|'
, DATE_FORMAT = 'yyyy-MM-dd HH:mm:ss'
));
```

The external file format specifies that the external files (uploaded in *step 1*) are pipe-delimited text files. The format types supported are delimited text, Hive RCFile, Hive ORC, and Parquet.

10. Execute the following query to create an external table:

```
CREATE EXTERNAL TABLE ext_orders(
InvoiceNo varchar(100),
StockCode varchar(100),
Description varchar(1000),
Quantity int,
InvoiceDate varchar(100),
UnitPrice decimal(10,2),
CustomerID varchar(500),
Country varchar(500),
orderid int
)
WITH
(
LOCATION='/orders/'
, DATA_SOURCE = adls
, FILE_FORMAT = orderFileFormat
, REJECT_TYPE = VALUE
, REJECT_VALUE = 0
);
GO
```

The location parameter specifies the directory that has the files to be read. The directory should exist in the Azure Data Lake Storage account as created in *step 1*. Execute the following query to count the number of rows in the external table created in *step 10*:

```
Select count(*) from ext_orders;
```

The result is shown in the following screenshot:

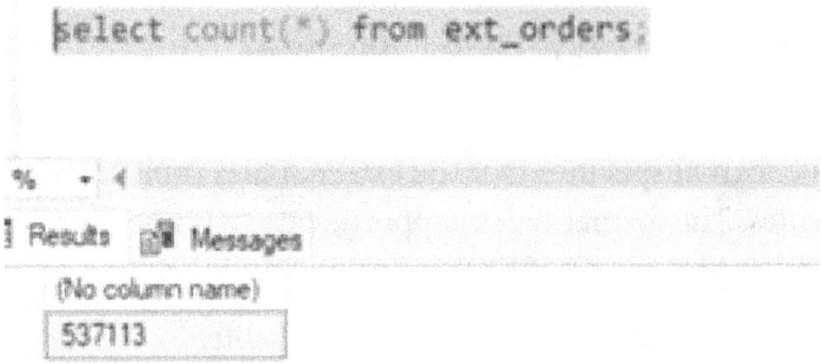

Figure 3.15 – Getting the number of rows in the external table

The data is read directly from the text files in the Azure Data Lake Storage account directory. The text files in the Azure Data Lake Storage account directory should be of the same schema as defined by the external table.

> **Note**
> If you get an error running the preceding query, make sure that the text files uploaded to the Azure Data Lake Storage account in *step 1* are UTF-8-encoded. If not, save the files as UTF-8 on your local machine and re-run *step 1* to upload the files to Azure Data Lake Storage.

11. Execute the following query to load the data from the external table into a table in the Synapse SQL pool:

```
SELECT * into Orders FROM ext_orders
OPTION (LABEL = 'CTAS : Load orders');
```

> **Note**
> We can also use the CREATE TABLE AS command to load the data into the orders table. We'll cover this later in the chapter.

How it works...

We start by creating a new Azure Data Lake Gen2 Storage account. To do this, we use the New-AzStorageAccount PowerShell command. We provide the resource group name, storage account name, SkuName, location, kind, and access tier. We set EnableHierarchicalNamespace to True to create a Data Lake Gen2 account and not an Azure Blob storage account. We store the storage account object in a variable to be used in later commands.

We create a new filesystem or container in the storage account using the `New-AzDatalakeGen2FileSystem` command. We provide the filesystem name and the Data Lake Storage account context.

We create a new directory inside the filesystem using the `New-AzDataLakeGen2Item` command. We specify the filesystem name, storage account context, directory path, and switch directory to specify that the item to be created is a directory.

We upload the data files in the directory. We use the `New-AzDataLakeGen2Item` command to upload the files. We provide the filesystem, the source (the file path in the local system), the Data Lake Storage account context, and the path. The `path` parameter is of the form `"<directory>/filename.extension"`. If we have to upload a file, say in the `sales/2012` directory, then the `path` parameter value will be `sales/2012/orders.txt`

We use a `foreach` loop to iterate through all of the files in the folder and upload them to Data Lake Storage.

To read and load the data from the files uploaded to Data Lake Storage, we need to define the *external data source*. The external data source is the Data Lake Storage account. We create an *external table* that points to the orders directory in the e-commerce filesystem.

The external table allows us to use T-SQL to query the CSV files directly into SSMS.

We can then transform or load or directly load the data from the external table into a table in the Synapse SQL pool.

Loading data into a SQL pool using the COPY INTO statement

The `COPY INTO` statement provides a faster and easier way to bulk insert data from Azure storage. We can use one `T-SQL` `COPY INTO` statement to ingest data instead of creating multiple database objects. At the time of writing this book, the `COPY INTO` statement is in preview.

In this recipe, we'll use the `COPY INTO` statement to load data into an Azure Synapse SQL pool.

Getting ready

Before you start, log in to Azure from PowerShell. To do this, execute the following command and follow the instructions to log in to Azure:

```
Connect-AzAccount
```

You need a Synapse SQL pool to perform the steps in the recipe. If you don't have an existing Synapse SQL pool, you can create one using the steps from the *Provisioning and connecting to an Azure Synapse SQL pool using PowerShell* recipe.

How to do it...

Follow the given steps to import data into a Synapse SQL pool from Azure Data Lake Storage Gen2:

1. Follow *step 1* in the *Loading data into a SQL pool using PolyBase with T-SQL* recipe to create an Azure Data Lake Storage Gen2 account and upload files to the account.

2. Open SSMS and connect to the Azure Synapse SQL pool in Object Explorer:

Figure 3.16 – Connecting to the Synapse SQL pool from SSMS

3. From the **Object Explorer** panel, expand **Databases** and right-click the **azadesqldw** SQL pool. From the right-click menu, select **New Query**:

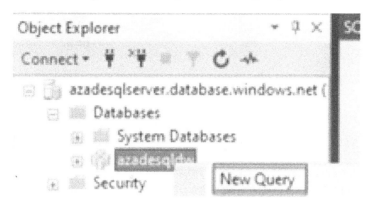

Figure 3.17 – Opening a new query window in SSMS

4. Execute the following query in the new query window to create the `orders` table:

```
IF OBJECT_ID('Orders', 'U') IS NOT NULL
   DROP TABLE Orders;
GO
CREATE TABLE Orders(
InvoiceNo varchar(100),
StockCode varchar(100),
Description varchar(1000),
Quantity int,
InvoiceDate varchar(100),
UnitPrice decimal(10,2),
CustomerID varchar(500),
Country varchar(500),
orderid int
)
WITH
(
DISTRIBUTION = ROUND_ROBIN,
CLUSTERED COLUMNSTORE INDEX
);
```

5. Execute the `COPY INTO` command to insert the data from Data Lake Storage into the `orders` table:

```
COPY INTO orders
from 'https://azadedls.dfs.core.windows.net/ecommerce/
orders/*.txt'
with
(
    fieldterminator = '|'
    ,encoding='utf8'
    ,CREDENTIAL=(IDENTITY= 'storage account key',
SECRET='RmM0cmNi0SUZ1OP3yy6HW61xg9KvjWW1Mui9ET/
1f41duSi8SsVItYtwiwvH4R7aku0DGY/KFG6qW/p5gHocvQ==')
)
```

6. Execute the following query to count the number of rows in the `orders` table:

```
Select count(*) from orders;
```

Here's the result:

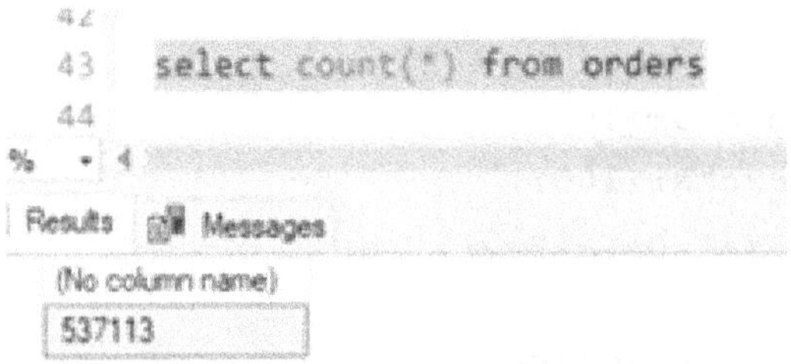

Figure 3.18 – Getting the number of records in the orders table

How it works...

The `COPY INTO` statement provides a faster and easier way to ingest data into a Synapse SQL pool. We can use one single T-SQL command without the need to create multiple objects to ingest data.

The `COPY INTO` statement provides the following important benefits over external tables:

- Requires just a single query to import files.

- Uses a shared access signature instead of a storage account key to access the storage account. This provides added security as the shared access signature has limited access to the blobs compared to the storage account key, which provides full access to the storage account.

- We use wildcards or specify multiple files when importing data. For example, we can specify `*.txt` to import only `.txt` files or `*.csv` to import `.csv` files.

- Specifies a separate location for error files.

- Uses SQL Server data formats for `.csv` files.

Implementing workload management in an Azure Synapse SQL pool

A data warehouse usually has mixed workloads, such as data import or data warehouse updates, reporting queries, aggregation queries, and data export. Running all of these queries in parallel results in resource challenges in the data warehouse.

Workload management uses workload classification, workload importance, and isolation to provide better control over how the workload uses the system resources.

In this recipe, we'll learn how to prioritize a workload using Windows classifiers and importance.

Getting ready

Before you start, open SSMS and log in to Azure SQL Server.

You need a Synapse SQL pool to perform the steps in this recipe. If you don't have an existing Synapse SQL pool, you can create one using the steps from the *Provisioning and connecting to an Azure Synapse SQL pool using PowerShell* recipe.

The recipe requires the `orders` table in the Synapse SQL pool. If you don't have it, follow the *Loading data into a SQL pool using the COPY INTO statement* recipe to create and load the `orders` table.

How to do it...

We can classify a workload by creating a workload classifier on five parameters: member name or username, label, context, start time, and end time. Each of these parameters has a numerical weightage. The sum of the weight of the parameter determines the query execution order.

In this recipe, we'll look at multiple scenarios and understand how to use the parameters to classify the workload.

Classifying a workload based on member name

Say we manage a data warehouse for an e-commerce store. There's a sales report that runs daily and is used by management to decide on the sales strategy.

The data warehouse is also used by the developers and analysts to run ad hoc queries for reporting or analytics. However, the sales report is of high importance and should not wait or get blocked because of ad hoc queries.

Follow the given steps to create a workload classifier to implement the preceding requirements:

1. Open a new query window in SSMS and execute the following query against the master database:

    ```
    CREATE LOGIN salesreport WITH PASSWORD='Dwh@1234'
    GO
    CREATE LOGIN adhocquery WITH PASSWORD ='Dwh@1234'
    ```

    The preceding query creates two logins: salesreport and adhocquery. The salesreport login is to run the daily sales report and the adhocquery login is for the analyst to run ad hoc queries.

2. Open a new query window in SSMS and execute the following query against the Synapse SQL pool, azadesqldw:

    ```
    CREATE USER salesreport FROM LOGIN salesreport
    GO
    CREATE USER adhocquery FROM LOGIN adhocquery
    GO
    GRANT SELECT ON orders TO adhocquery
    GO
    GRANT SELECT ON orders TO salesreport
    ```

The preceding query creates users for the `salesreport` and `adhocquery` logins in the `azadesqldw` SQL pool and grants read access to the `orders` table.

3. Open a new query window in SSMS and connect to the `azadesqldw` Synapse SQL pool. Execute the following T-SQL query:

```
CREATE WORKLOAD CLASSIFIER [wlcsalesreport]
WITH (WORKLOAD_GROUP = 'xlargerc'
,MEMBERNAME = 'salesreport'
,IMPORTANCE = HIGH);
GO
CREATE WORKLOAD CLASSIFIER [wlcadhocquery]
WITH (WORKLOAD_GROUP = 'xlargerc'
,MEMBERNAME = 'adhocquery'
,IMPORTANCE = LOW);
```

> **Note**
>
> The `workload_group` class can be either an existing resource class or a custom workload group. To find out more about workload groups, refer to the documentation at `https://docs.microsoft.com/en-us/sql/t-sql/statements/create-workload-group-transact-sql?toc=/azure/synapse-analytics/sql-data-warehouse/toc.json&bc=/azure/synapse-analytics/sql-data-warehouse/breadcrumb/toc.json&view=azure-sqldw-latest`.

The preceding query creates two Windows classifiers:

a) `wlcsalesreport` for the `salesreport` user with the `xlargerc` workload group and importance as `high`

b) `wlcadhocquery` for the `adhocquery` user with the `xlargerc` workload group and importance as `low`

4. Open a new query window in SSMS and connect to the `azadesqldw` SQL pool. Copy and paste the following T-SQL into the query window:

```
SELECT
r.session_id,
s.login_name,
r.status,
r.resource_class,
```

```
    r.classifier_name,
    r.group_name,
    r.importance,
    r.submit_time,
    r.start_time
FROM sys.dm_pdw_exec_requests r join sys.dm_pdw_exec_
sessions s
on s.session_id=r.session_id
WHERE
r.status in ('Running','Suspended')
   AND s.session_id <> session_id()
   AND s.login_name in ('salesreport','adhocquery')
ORDER BY submit_time
```

The preceding query is to monitor the queries running on the Synapse SQL pool.

5. Open a new PowerShell command window and execute the following command to run the workload:

```
.\ADECookbook\Chapter03\Scripts\ExecuteWorkload.
ps1 -azuresqlserver azadesqlserver -sqldw azadesqldw
-adhocquerypath C:\ADECookbook\Chapter03\Scripts\
adhocquery.sql -salesreportpath C:\ADECookbook\Chapter03\
Scripts\salesreport.sql -sqlcmdpath "C:\Program Files\
Microsoft SQL Server\Client SDK\ODBC\170\Tools\Binn\
sqlcmd.exe"
```

Note

You'll have to change the parameter values before running the command as they may differ in your case.

The preceding command executes three queries with a delay of 5 seconds within each query execution. The first two queries are the adhoc query from the adhocquery user and the third query is run by the salesreport user.

While the queries are running, quickly switch to the query window mentioned in *step 4* and run the monitoring query:

```
32   SELECT
33       r.session_id,
34       s.login_name,
35       r.status,
36       r.resource_class,
37       r.classifier_name,
38       r.group_name,
39       r.importance,
40       r.submit_time,
41       r.start_time|
42   FROM sys.dm_pdw_exec_requests r join sys.dm_pdw_exec_sessions s
```

110 %

session_id	login_name	status	resource_class	classifier_name	group_name	importance	submit_time	start_time
1	SID111	adhocquery	Running	xlargerc	wlcadhocquery	xlargerc	low	2020-05-13 19:41:35.390	2020-05-13 19:41:35.483

Figure 3.19 – Monitoring query

Observe that there's a query from the `adhocquery` login. The query is using the `wlcadhocquery` classifier as we specified in *step 3*. The query's importance is `low`.

Re-run the monitoring query in 5 seconds:

```
32   SELECT
33       r.session_id,
34       s.login_name,
35       r.status,
36       r.resource_class,
37       r.classifier_name,
38       r.group_name,
39       r.importance,
40       r.submit_time,
41       r.start_time|
42   FROM sys.dm_pdw_exec_requests r join sys.dm_pdw_exec_sessions s
```

110 %

session_id	login_name	status	resource_class	classifier_name	group_name	importance	submit_time	start_time
1	SID111	adhocquery	Running	xlargerc	wlcadhocquery	xlargerc	low	2020-05-13 19:41:35.390	2020-05-13 19:41:35.483
2	SID112	adhocquery	Suspended	xlargerc	wlcadhocquery	xlargerc	low	2020-05-13 19:41:40.030	NULL
3	SID113	salesreport	Suspended	xlargerc	wlcsalesreport	xlargerc	high	2020-05-13 19:41:46.407	NULL

Figure 3.20 – Monitoring query

Observe that we now have three queries: two from the `adhocquery` user and one from the `salesreport` user. The `salesreport` query has its importance as `high` as it's using the `wlssalesreport` classifier, as specified in *step 3*.

The `adhocquery` user with `session_id` as `SID111` is running and the two other queries are waiting.

> **Note**
>
> The `azadesqldw` SQL pool is of the DW100c performance tier. This tier allows only one query at a time in the `xlargerc` resource class.

Re-run the monitoring query:

```
32    SELECT
33          r.session_id,
34          s.login_name,
35          r.status,
36          r.resource_class,
37          r.classifier_name,
38          r.group_name,
39          r.importance,
40          r.submit_time,
41          r.start_time
42    FROM sys.dm_pdw_exec_requests r join sys.dm_pdw_exec_sessions s
```

110 %

Results　Messages

	session_id	login_name	status	resource_class	classifier_name	group_name	importance	submit_time	start_time
1	SID112	adhocquery	Suspended	xlargerc	wlcadhocquery	xlargerc	low	2020-05-13 19:41:40.030	NULL
2	SID113	salesreport	Running	xlargerc	wlcsalesreport	xlargerc	high	2020-05-13 19:41:46.407	2020-05-13 19:42:22.297

Figure 3.21 – Monitoring query

Observe that the query with session ID `SID111` is complete. The `SID113` session is running even though it was submitted after the `SID112` session. This is because the `SID` session has an importance of `high` and is therefore picked up ahead of `SID112`.

Classifying a workload based on query label

In the previous scenario, we created a `salesreport` user to run the daily sales report. Consider a scenario where there are multiple sales reports, some of them run by developers and some of them run by managers using the `salesreport` member name. The reports run by managers are of more importance than the reports run by developers. Let's create a workload classifier to implement this requirement:

1. Open a new SSMS query window, connect to the `azadesqldw` SQL pool, and execute the following T-SQL command:

```
CREATE WORKLOAD CLASSIFIER [wlcsalesmgr]
WITH (WORKLOAD_GROUP = 'xlargerc'
,MEMBERNAME = 'salesreport'
```

```
,WLM_LABEL = 'manager'
,IMPORTANCE = HIGH);
GO
CREATE WORKLOAD CLASSIFIER [wlcsalesdev]
WITH (WORKLOAD_GROUP = 'xlargerc'
,MEMBERNAME = 'salesreport'
,WLM_LABEL = 'developer'
,IMPORTANCE = LOW);
GO
```

The preceding commands create two workload classifiers:

a) wlcsalesmgr, with its member name as salesreport, WLM_LABEL as manager, and the importance set as HIGH. This says that a query from the salesreport member with its query label set to manager is to be prioritized over queries from other users or queries from the same user with a different label.

b) wlcsalesdev, with its member name as salesreport, WLM_LABEL as developer, and the importance set as LOW. This says that a query from the salesreport member with its query label set to developer will be of low priority.

2. Open a PowerShell command window and execute the following command:

```
.\ADECookbook\Chapter03\Scripts\ExecuteWorkloadlabel.
ps1 -azuresqlserver azadesqlserver -sqldw azadesqldw
-salesdevpath C:\ADECookbook\Chapter03\Scripts\
salesreportdev.sql -salesmgrpath C:\ADECookbook\
Chapter03\Scripts\salesreportmgr.sql -sqlcmdpath "C:\
Program Files\Microsoft SQL Server\Client SDK\ODBC\170\
Tools\Binn\sqlcmd.exe"
```

> **Note**
> You may have to change the parameter values in the preceding command before execution.

The preceding command executes three queries with a delay of 5 seconds between each query execution. All three queries are executed by the salesreport user. However, the first two queries are labeled as developer and the third query is labeled as manager.

> **Note**
>
> We can label a query in a Synapse SQL pool by adding OPTION (Label = "label value") to the end of the query – for example, SELECT TOP 100 * FROM orders OPTION(LABEL = "top100orders").

3. While the workload is running, quickly run the following monitoring query in SSMS:

```
SELECT
        r.session_id,
        s.login_name,
        r.[label],
        r.status,
        r.resource_class,
        r.classifier_name,
        r.group_name,
        r.importance,
        r.submit_time,
        r.start_time
FROM sys.dm_pdw_exec_requests r join sys.dm_pdw_exec_
sessions s
on s.session_id=r.session_id
WHERE
r.status in ('Running','Suspended')
   AND s.session_id <> session_id()
   AND s.login_name = 'salesreport'
ORDER BY submit_time
```

You should get an output as shown in the following screenshot:

Figure 3.22 – Monitoring query

Observe that there are three queries in the result set. The SID116 and SID117 sessions are labeled as developer and are running under the wlcsalesdev classifier with their importance set to low. The SID118 session is labeled as manager and is running under the wlcsalesmgr classifier with its importance set to high.

The SID116 session is currently running as it was submitted ahead of the SID117 and SID118 sessions.

Re-run the monitoring query in another 5 seconds:

Figure 3.23 – Monitoring query

Observe that the SID118 session is running and the SID117 session is suspended even though it was submitted ahead of SID118. This is because the SID118 session has a label, manager. Therefore, it's mapped to the wlcsalesmgr classifier with its importance set to high.

Classifying a workload based on context

In the previous example, we saw how we can use query labels to prioritize workloads. However, consider a scenario where we have two queries with the same label. Usually, it doesn't matter which one is picked first; however, consider a scenario where we want to show a sales dashboard to a client and we want to make sure that the dashboard gets priority over any other query.

In this example, we'll learn how to prioritize queries using the `WLM_CONTEXT` classifier parameter:

1. Open a new SSMS query window, connect to the `azadelsqdw` Synapse SQL pool, and execute the following query:

```
CREATE WORKLOAD CLASSIFIER [wlcsalesdashboard]
WITH (WORKLOAD_GROUP = 'xlargerc'
,MEMBERNAME = 'salesreport'
,WLM_LABEL = 'manager'
,WLM_CONTEXT = 'dashboard'
,IMPORTANCE = HIGH);
GO


CREATE WORKLOAD CLASSIFIER [wlcsalesqueries]
WITH (WORKLOAD_GROUP = 'xlargerc'
,MEMBERNAME = 'salesreport'
,WLM_LABEL = 'manager'
,IMPORTANCE = LOW);
```

The preceding query creates two classifiers: `wlcsalesdashboard` and `wlcsalesqueries`. The two classifiers are the same except that the `wlcsalesdashboard` classifier specifies `WLM_CONTEXT` as `dashboard` and is of high importance.

2. Open a PowerShell command window and execute the following command:

```
.\ADECookbook\Chapter03\Scripts\ExecuteWorkloadcontext.
ps1 -azuresqlserver azadesqlserver -sqldw azadesqldw
-salecontextpath C:\ADECookbook\Chapter03\Scripts\
salesreportcontext.sql -salesnocontextpath C:\
ADECookbook\Chapter03\Scripts\salesreportnocontext.sql
-sqlcmdpath "C:\Program Files\Microsoft SQL Server\Client
SDK\ODBC\170\Tools\Binn\sqlcmd.exe"
```

> **Note**
>
> You'll have to change the parameter values before running the preceding command.

The preceding command executes three queries with a delay of 5 seconds within each query execution. All three queries are executed by the `salesreport` user and have a query label of `manager`. However, the third query, `C:\ADECookbook\Chapter03\Scripts\salesreportcontext.`, specifies the session context as `dashboard`.

> **Note**
>
> The session context can be set using the `sys.sp_set_session_context` stored procedure. For example, to set the session context as `dashboard`, execute the `EXEC sys.sp_set_session_context @key = 'wlm_context', @value = 'dashboard'` query.

3. While the workload is running, quickly execute the following monitoring query in SSMS against the `azadesqldw` database:

```
SELECT
        r.session_id,
        s.login_name,
        r.[label],
        r.status,
        r.resource_class,
        r.classifier_name,
        r.group_name,
        r.importance,
        r.submit_time,
        r.start_time
```

```
FROM sys.dm_pdw_exec_requests r join sys.dm_pdw_exec_
sessions s
on s.session_id=r.session_id
WHERE
r.status in ('Running','Suspended')
   AND s.session_id <> session_id()
   AND s.login_name = 'salesreport'
   and r.classifier_name is not null
ORDER BY submit_time
```

You should get an output as shown in the following screenshot:

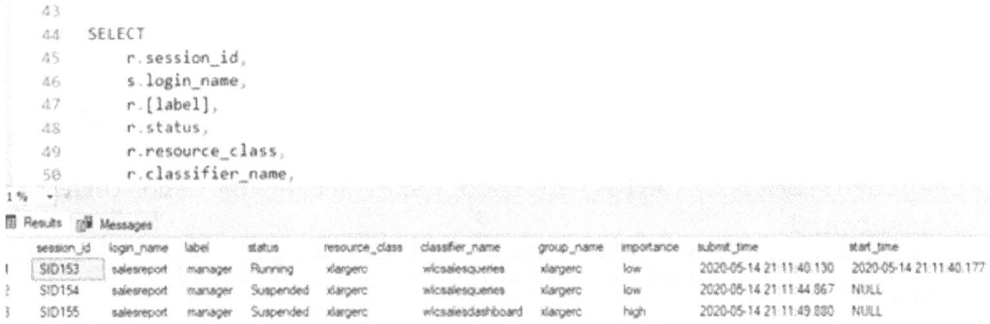

Figure 3.24 – Monitoring query

Observe that there are three queries in the result set. The SID153 and SID154 sessions are running under wlcsalesqueries with their importance set to low and the SID155 session is running under the wlcsalesdashboard classifier with its importance set to high. This is because the SID155 session explicitly sets WLM_CONTEXT to dashboard.

Re-run the monitoring query in another 5–8 seconds:

```
44    SELECT
45        r.session_id,
46        s.login_name,
47        r.[label],
48        r.status,
49        r.resource_class,
50        r.classifier_name,
51        r.group_name,
52        r.importance,
```

Figure 3.25 – Monitoring query

Observe that once the `SID153` session completes, the `SID155` session is picked up for execution even though it was submitted 5 seconds later than the `SID154` session.

This is because it's running under the `wlcsalesdashboard` classifier with `high` importance.

How it works...

In this series of examples, we learned how to classify a workload using the `workload` classifier. The `workload` classifier uses five parameters to classify a workload. Each parameter has a weighting value. The following are the parameters with their weighting values:

Parameter	Weight
USER	64
ROLE	32
WLM_LABEL	16
WLM_CONTEXT	8
START TIME AND ENDTIME	4

Figure 3.26 – Parameters with their weighting values

The workload classifier is applied by adding up the weights across all parameters. Consider the scenario from the *Classifying a workload based on context* recipe.

There are two queries, salesreportcontext.sql and salesreportnocontext. sql. The query text is given as follows:

- salesreportcontext.sql:

```
-- salesreportcontext.sql
EXEC sys.sp_set_session_context @key = 'wlm_context', @
value = 'dashboard'
select
Datepart(year,invoicedate) as [year],
Datepart(month,invoicedate) as [month],
stockcode,
customerid,
country,
sum(quantity*unitprice) as totalsales
from orders
group by
Datepart(year,invoicedate),Datepart(month,invoicedate),
country,stockcode,customerid
order by country,[year],[month],stockcode,customerid
option(label = 'manager')
```

- salesreportnocontext.sql:

```
-- salesreportnocontext.sql
select
Datepart(year,invoicedate) as [year],
Datepart(month,invoicedate) as [month],
stockcode,
customerid,
country,
sum(quantity*unitprice) as totalsales
from orders
group by
Datepart(year,invoicedate),Datepart(month,invoicedate),
country,stockcode,customerid
```

```
order by country, [year], [month], stockcode, customerid
option(label = 'manager')
```

Observe that the difference between the two queries is that in `salesreportcontext.sql`, we set the `WLM_CONTEXT` session parameter to `dashboard`. Therefore, as per the `wlcsalesdashboard` classifier, the query has a total weight of `(MEMBERNAME) 64 + (WLM_LABEL) 16 + (WLM_CONTEXT) 8 = 88`. The `salesreportnocontext.sql` query falls into the `wlcsalesqueries` classifier group as there's no context specified. It has a total weight of `(MEMBERNAME) 64 + (WLM_LABEL) 16 = 80`. Therefore, the `salesreportcontext.sql` query is of high importance and is picked up ahead of `salesreportnocontext.sql` for execution.

Optimizing queries using materialized views in Azure Synapse Analytics

Views are an old concept in SQL Server and are often used to encapsulate complex queries into virtual tables. We can then replace the query with the virtual table wherever required. A standard view is just a name given to the complex query. Whenever we query the standard view, it accesses the underlying tables in the query to fetch the result set.

Materialized views, unlike standard views, maintain the data as a physical table instead of a virtual table. The view data is maintained just like a physical table and is refreshed automatically whenever the underlying tables are updated.

In this recipe, we'll learn how to optimize queries using materialized views.

Getting ready

Before you start, open SSMS and log in to Azure SQL Server.

You need a Synapse SQL pool to perform the steps in this recipe. If you don't have an existing Synapse SQL pool, you can create one using the steps from the *Provisioning and connecting to an Azure Synapse SQL pool using PowerShell* recipe.

The recipe requires the `orders` table in the Synapse SQL pool. If you don't have it, follow the *Loading data into a SQL pool using the COPY INTO statement* recipe to create and load the `orders` table.

How to do it...

To optimize queries using materialized views, follow the given steps:

1. Open SSMS and connect to the Azure Synapse SQL pool in Object Explorer:

Figure 3.27 – Connecting to the Synapse SQL pool from SSMS

2. From the **Object Explorer** panel, expand **Databases** and right-click the **azadesqldw** SQL pool. From the right-click menu, select **New Query**:

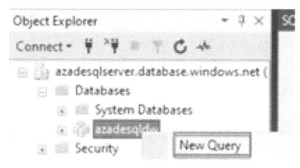

Figure 3.28 – Opening a new query window in SSMS

3. Copy and paste the following query in the query window:

```
SELECT
        country,
        sum(quantity*unitprice) as totalsales,
        count(*) as ordercount
FROM orders
GROUP BY country
ORDER BY country
```

4. Press *Ctrl + L* to display the estimated execution plan for the query:

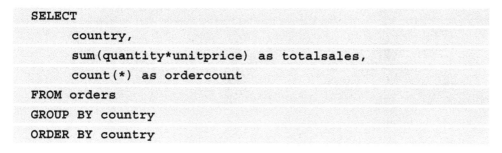

Figure 3.29 – Estimated execution plan

5.   Right-click on the **Get** operator and select **Properties**:

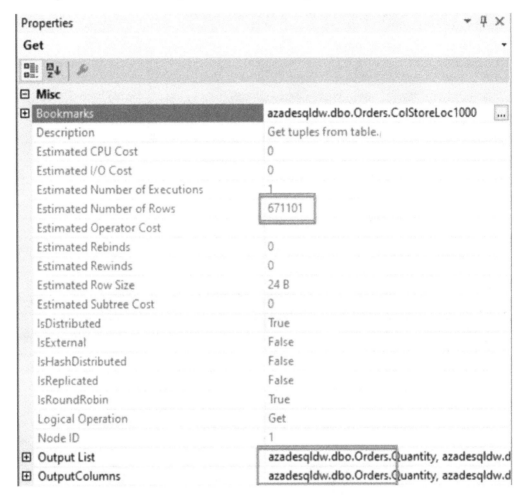

Figure 3.30 – Getting the operator properties

Observe that the query is getting the 671101 rows from the orders table. Let's create a materialized view and then execute the query again.

6. Execute the following query in an SSMS query window to create the materialized view:

```
CREATE materialized view mvw_salesreport
WITH (DISTRIBUTION=HASH(country))
AS
SELECT
      country,
      sum(quantity*unitprice) as totalsales,
      count(*) as ordercount
FROM dbo.orders
GROUP BY country
```

The preceding query creates the materialized view with a hash distribution on the country column on the query specified in *step 3*.

7. Press *Ctrl + L* to fetch the estimated execution plan of the query in *step 3*:

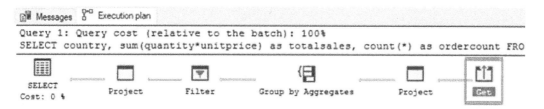

Figure 3.31 – Estimated execution plan

8. Right-click on the **Get** operator and select **Properties**:

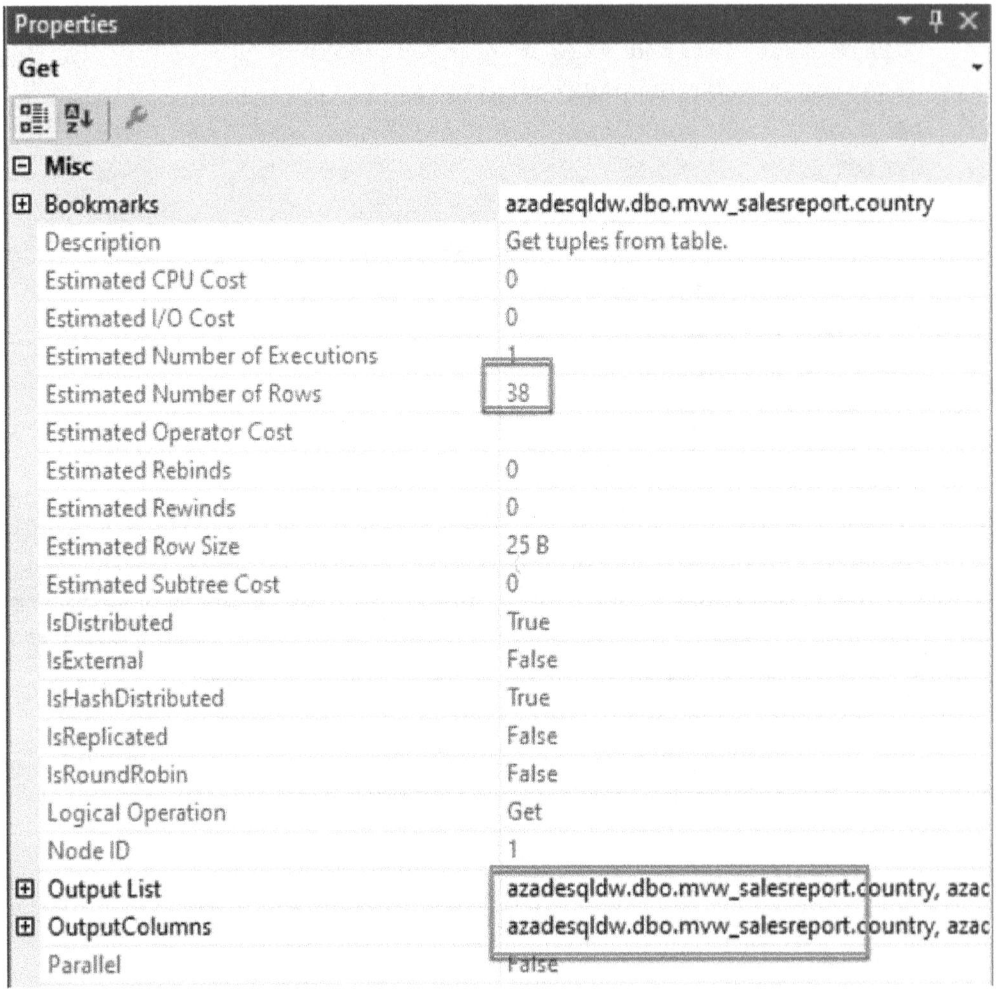

Figure 3.32 – Getting the operator properties

Observe that instead of scanning the orders table, the same query now gets the 38 rows from the materialized view, mvw_salesreport.

The database is intelligent enough to detect a supporting materialized view and uses it to improve the query performance. We can get better performance without changing the query.

How it works...

Materialized views are used to optimize aggregate queries. We can create a materialized view on aggregate queries and store the aggregations physically like any other physical table. As the aggregates are stored physically, they aren't recalculated every time the query runs and the query is therefore faster.

The materialized views are automatically refreshed as and when the data in the base table changes.

The data in a materialized view can have a different distribution than that of the base table. The orders table used in the preceding examples has round-robin data distribution. The round-robin distribution helps with faster data insertion. The materialized view has a hash distribution on the country column. The hash distribution supports faster join and aggregate queries.

The pdw_showmaterializedviewoverhead **database console command** (**DBCC**) displays the number of changes in base tables that are not yet applied to the materialized view. It is displayed as overhead_ratio.

If the overhead ratio is too high, we may see performance degradation and we should rebuild the materialized view.

To rebuild a materialized view, we can use the following query:

```
ALTER MATERIALIZED VIEW dbo.mvw_salesreport REBUILD
```

4

Control Flow Activities in Azure Data Factory

Azure Data Factory (**ADF**) is a cloud extract, transform, and load service. ADF is used to read data from multiple sources, transforming and loading the data into databases to be consumed for reporting, data science, or machine learning. ADF provides a code-free UI to develop, manage, and maintain data pipelines.

ADF is compatible with **SQL Server Integration Services** (**SSIS**). Existing SSIS packages can be easily migrated and run in ADF.

ADF is an essential Azure data engineering product and is a must for an Azure data engineer to learn.

In this chapter, we'll cover the following recipes:

- Implementing control flow activities
- Implementing control flow activities – Lookup and If activities
- Triggering a pipeline in Azure Data Factory

Technical requirements

For this chapter, the following are required:

- A Microsoft Azure subscription
- PowerShell 7
- Microsoft Azure PowerShell
- SQL Server Management Studio or Azure Data Studio

Implementing control flow activities

In this recipe, we'll learn how to implement common control flow activities such as Get Metadata, Filter, ForEach, and Stored Procedure. Control flows help in building complex iterative and conditional logic to support data ingestion.

In this recipe, we'll implement the following scenario using control flow activities.

We have a data lake directory that contains multiple file formats, such as .txt, .csv, and .tif. We need to copy .txt files from the directory to another directory. We'll also maintain a log of the file copy operation in an Azure SQL Database table.

Getting ready

Before you start, do the following:

1. Log into Azure from PowerShell. To do this, execute the following command and follow the instructions to log into Azure:

   ```
   Connect-AzAccount
   ```

2. Open https://portal.azure.com and log in using your Azure credentials.

How to do it...

Let's get started by creating the Azure SQL database and the required database objects to support the logging:

1. Execute the following command to provision a new Azure SQL database:

   ```
   .\ADE\azure-data-engineering-cookbook\Chapter04\5_
   ProvisioningAzureSQLDB.ps1 -resourcegroup packtade
   -servername azadesqlserver -databasename azadesqldb
   -password Sql@Server@1234 -location centralus
   ```

The preceding command creates the resource group if it doesn't exist, creates a new Azure SQL Server, Azure SQL Database, adds the host public IP address in the Azure SQL Server firewall, and allows Azure SQL Server access to all Azure services.

2. Once the database is provisioned, create the `FileStatusCopy` table and the `usp_UpateFileStatusCopy` stored procedure. To do this, execute the query in the `~\azure-data-engineering-cookbook\Chapter04\SQL\DatabaseObjects.sql` file.

    > **Note**
    >
    > You can run the query by using SQLCMD, SQL Server Management Studio, or the Azure portal query editor.

3. Execute the following command to create an Azure data factory:

    ```
    .\ADE\azure-data-engineering-cookbook\Chapter04\3_
    CreatingAzureDataFactory.ps1 -resourcegroupname packtade
    -location centralus -datafactoryname packtdatafactory
    ```

    > **Note**
    >
    > If you have an existing data factory, you can skip this step.

4. We'll now create a linked service to connect to Azure SQL Database. We'll use the definition file to create the linked service. To do that, open the `~\azure-data-engineering-cookbook\Chapter04\AzureSQLDBLinkedServiceDefinitionFile.txt` file and replace the data source, initial catalog, user ID, and password in the `connectionString` property.

5. Save and close the file. Then, execute the following command to create the linked service:

    ```
    Set-AzDataFactoryV2LinkedService -Name
    AzureSqlDatabaseLinkedService -DefinitionFile C:\
    ade\azure-data-engineering-cookbook\Chapter04\
    AzureSQLDBLinkedServiceDefinitionFile.txt
    -DataFactoryName packtdatafactory -ResourceGroupName
    packtade
    ```

6. Execute the following command to create a new Azure Date Lake Storage Gen2 account and upload the files:

```
.\ADE\azure-data-engineering-cookbook\Chapter04\1_
UploadOrderstoDataLake.ps1 -resourcegroupname packtade
-storageaccountname packtdatalakestore -location
centralus -directory C:\ADE\azure-data-engineering-
cookbook\Chapter04\Data\
```

The preceding command creates an Azure Data Lake Storage Gen2 account. It creates an `ecommerce` file system in the Azure Data Lake Storage Gen2 account and two directories, `orders` and `orderstxt`, in the `ecommerce` file system. It uploads all the files from the `Data` folder to the `orders` directory. You should get an output as shown in the following screenshot:

```
    Container Uri: https://packtdatalakestore.blob.core.windows.net/ecommerce

Name                   BlobType   Length      ContentType                LastModified            AccessTier Snapsh
                                                                                                            ime
----                   --------   ------      -----------                ------------            ---------- ------
orders                 BlockBlob  0           application/octet-stream   2020-06-14 13:57:55Z Hot
orders/orders1.txt     BlockBlob  12601902    application/octet-stream   2020-06-14 13:58:54Z Hot
orders/orders2.tif     BlockBlob  12758599    application/octet-stream   2020-06-14 13:59:19Z Hot
orders/orders2.txt     BlockBlob  12735799    application/octet-stream   2020-06-14 13:59:53Z Hot
orders/orders3.tif     BlockBlob  12758599    application/octet-stream   2020-06-14 14:00:27Z Hot
orders/orders3.txt     BlockBlob  12758599    application/octet-stream   2020-06-14 14:01:01Z Hot
orders/orders4.txt     BlockBlob  12886718    application/octet-stream   2020-06-14 14:01:30Z Hot
orders/orders5.txt     BlockBlob  1379322     application/octet-stream   2020-06-14 14:01:32Z Hot
orderstxt              BlockBlob  0           application/octet-stream   2020-06-14 13:57:56Z Hot


KeyName Value                                                                                  Permissions
------- -----                                                                                  -----------
key1    EvG77HNayTToXPkxPMPAr7iYVjS8wnGvdzbeUbiwKA90apg2+3BFxh3sIIj3tw+O2vUG+AwQRfVPCFvAFliDqQ==   Full
key2    gWREzLUmdDETWpBbRrT7DB2YnlbeVQ/iALkEKark2StaRKxzGr8VlyP4msgINLQzcoVTU42jBk3Px+Z0eT957g==   Full

PS C:\>
```

Figure 4.1 – Uploading files to Azure Data Lake Storage Gen2

Note the value for `key1`. It'll be used in the next step to create the linked service.

7. We'll now create a linked service to Azure Data Lake Storage Gen2.
    Open the `azure-data-engineering-cookbook\Chapter04\`
    `DataLakeStorageLinkedServiceDefinitionFile.txt` definition file.
    Then, replace the value of `AccountKey` with the one noted in the previous step
    (`key1`). You may also have to change the Azure Data Lake Storage Gen2 account
    name in the URL. Save and close the file.

8. Execute the following command to create the linked service to connect to Azure Data Lake Storage Gen2:

```
Set-AzDataFactoryV2LinkedService -Name
DataLakeStoreLinkedService -DefinitionFile C:\
ade\azure-data-engineering-cookbook\Chapter04\
DataLakeStorageLinkedServiceDefinitionFile.
txt -ResourceGroupName packtade -DataFactoryName
packtdatafactory
```

> **Note**
>
> You may have to replace the data factory name and resource group name in the preceding command.

9. We'll now create the required datasets. We need three datasets: one for `orders` (the source data lake directory), one for `orderstxt` (the destination data lake directory), and one dataset with a file path that includes a single file. We'll create the dataset by providing the definition file to the PowerShell command.

10. To create the `orders` dataset, open the `~\azure-data-engineering-cookbook\Chapter04\ordersdatasetdefinitionfile.txt` file. Modify the linked service name if it's different from what we created in *step 4*. Modify the `fileSystem` and `folderPath` values if they are different from the ones created in *step 3*:

```
{
    "name": "orders",
    "properties": {
        "linkedServiceName": {
            "referenceName": "DataLakeStoreLinkedService",
            "type": "LinkedServiceReference"
        },
        "annotations": [],
        "type": "DelimitedText",
        "typeProperties": {
            "location": {
                "type": "AzureBlobFSLocation",
```

```
        "folderPath": "orders",
        "fileSystem": "ecommerce"
      },
      "columnDelimiter": ",",
      "escapeChar": "\\",
      "quoteChar": "\""
    },
    "schema": []
  }
}
```

Save and close the file.

11. Execute the following PowerShell command to create the `orders` dataset:

```
Set-AzDataFactoryV2Dataset -Name orders -DefinitionFile
C:\ade\azure-data-engineering-cookbook\Chapter04\
OrdersDatasetDefinitionfile.txt -ResourceGroupName
packtade -DataFactoryName packtdatafactory
```

> **Note**
> You may have to change the resource group and data factory name in the
> preceding command.

12. To create the `orderstxt` dataset, open the `~\azure-data-engineering-cookbook\Chapter04\OrderstxtDatasetDefinitionfile.txt` definition file and modify the linked service name if it's different from what we created in *step 4*. Modify the `fileSystem` and `folderPath` values if they are different from the ones created in *step 3*:

```
{
  "name": "orderstxt",
  "properties": {
    "linkedServiceName": {
      "referenceName": "DataLakeStoreLinkedService",
      "type": "LinkedServiceReference"
```

```
    },
    "annotations": [],
    "type": "DelimitedText",
    "typeProperties": {
      "location": {
        "type": "AzureBlobFSLocation",
        "folderPath": "orderstxt",
        "fileSystem": "ecommerce"
      },
      "columnDelimiter": ",",
      "escapeChar": "\\",
       "quoteChar": "\""
    },
    "schema": []
  }
}
```

fileSystem refers to the Azure Data Lake Gen2 container created in *step 6*. folderPath refers to the folder in the container or the file system that contains the orders file.

Save and close the file.

13. Execute the following PowerShell command to create the orderstxt dataset:

```
Set-AzDataFactoryV2Dataset -Name orderstxt
-DefinitionFile C:\ade\azure-data-engineering-
cookbook\Chapter04\OrderstxtDatasetDefinitionfile.
txt -ResourceGroupName packtade -DataFactoryName
packtdatafactory
```

> **Note**
>
> You may have to change the resource group and data factory name in the preceding command.

14. The third and last dataset we'll create is ordersfile. To create the dataset, open the ~\azure-data-engineering-cookbook\Chapter04\ OrdersfileDatasetDefinitionfile.txt file and modify the linked service name if it's different from what we created in *step 4*. Modify the fileSystem and folderPath values if they are different from the ones created in *step 3*:

```
{
    "name": "ordersfile",
    "properties": {
        "linkedServiceName": {
            "referenceName": "DataLakeStoreLinkedService",
            "type": "LinkedServiceReference"
        },
        "annotations": [],
        "type": "DelimitedText",
        "typeProperties": {
            "location": {
                "type": "AzureBlobFSLocation",
                "fileName": "orders1.txt",
                "folderPath": "orders",
                "fileSystem": "ecommerce"
            },
            "columnDelimiter": "|",
            "escapeChar": "\\",
            "quoteChar": "\""
        },
        "schema": [
            {
                "type": "String"
            }
        ]
    }
}
```

Save and close the file.

15. Execute the following PowerShell command to create the `ordersfile` dataset:

```
Set-AzDataFactoryV2Dataset -Name ordersfile
-DefinitionFile C:\ade\azure-data-engineering-
cookbook\Chapter04\OrdersfileDatasetDefinitionfile.
txt -ResourceGroupName packtade -DataFactoryName
packtdatafactory
```

> **Note**
>
> You may have to change the resource group and data factory name in the
> preceding command.

16. To list out the created datasets, execute the following PowerShell command:

```
$df = Get-AzDataFactoryV2 -ResourceGroupName packtade
-Name packtdatafactory
Get-AzDataFactoryV2Dataset -DataFactory $df
```

You should get an output as shown in the following screenshot:

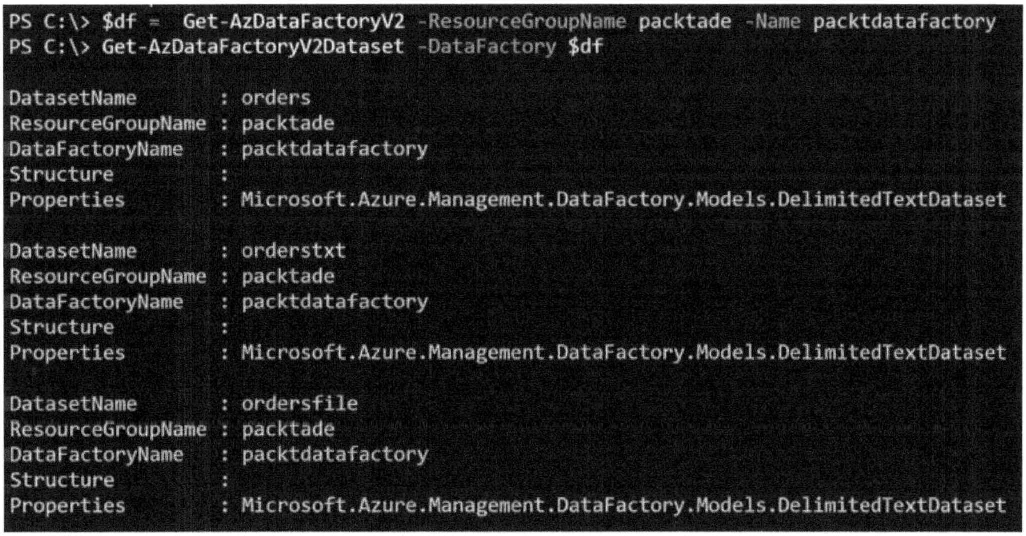

```
PS C:\> $df = Get-AzDataFactoryV2 -ResourceGroupName packtade -Name packtdatafactory
PS C:\> Get-AzDataFactoryV2Dataset -DataFactory $df

DatasetName        : orders
ResourceGroupName  : packtade
DataFactoryName    : packtdatafactory
Structure          :
Properties         : Microsoft.Azure.Management.DataFactory.Models.DelimitedTextDataset

DatasetName        : orderstxt
ResourceGroupName  : packtade
DataFactoryName    : packtdatafactory
Structure          :
Properties         : Microsoft.Azure.Management.DataFactory.Models.DelimitedTextDataset

DatasetName        : ordersfile
ResourceGroupName  : packtade
DataFactoryName    : packtdatafactory
Structure          :
Properties         : Microsoft.Azure.Management.DataFactory.Models.DelimitedTextDataset
```

Figure 4.2 – Listing datasets

We'll now start working on the pipeline.

17. In the Azure portal, under **All resources**, find and open `packtdatafactory`. Select **Monitor & Author** and then select **Create Pipeline**. Name the pipeline `Pipeline-ControlFlow-Activities`.

18. Under the **Activities** tab, expand **General** and drag and drop the **Get Metadata** activity. Rename the activity `Get Child Items`.

19. Switch to the **Dataset** tab and select the `orders` dataset from the **Dataset** dropdown.

20. Under the **Field** list heading, click **New** and select **Child Items** under the **ARGUMENT** heading:

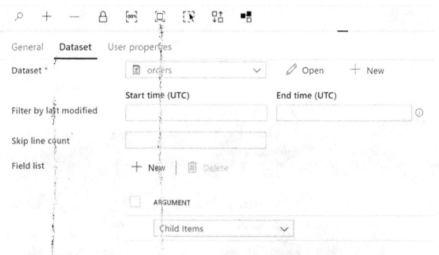

Figure 4.3 – Configuring the Get Metadata activity

The **Get Metadata** activity will get all of the files in the `orders` dataset and will return the filenames.

21. The next step is to filter `.txt` files. To do this, drag and drop the **Filter** activity from the **Iterations & conditionals** tab. Click on the small green box in the **Get Metadata** activity and drag it to the **Filter** activity. This will cause the **Filter** activity to start when the **Get Metadata** activity completes.

22. Select the **Filter** activity. Under the **General** tab, rename it `Filter text files`. Then, switch to the **Settings** tab:

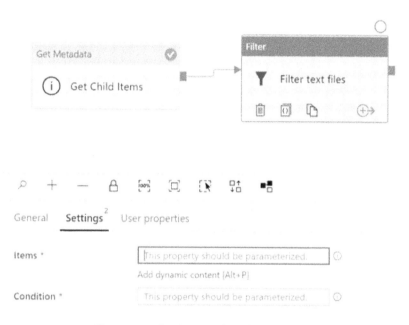

Figure 4.4 – Configuring the Filter activity

The **Filter** activity takes two parameters, **Items** and **Condition**. The **Items** parameter is an array on which the filter condition, as specified in the **Condition** parameter, is specified.

23. Click on the **Items** text box, and then click on the **Add dynamic content** link.

24. In the **Add dynamic content** dialog, scroll to the end. Under **Activity outputs**, select **Get Child Items**:

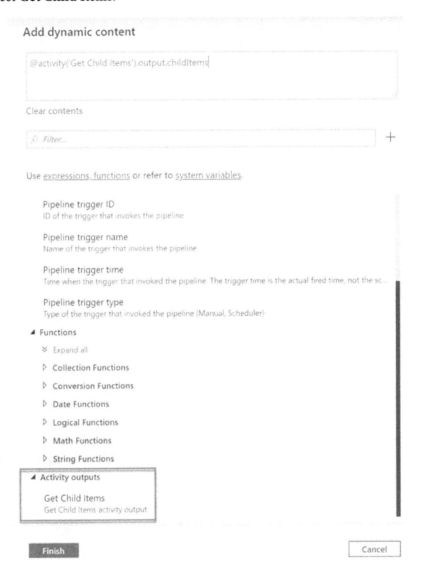

Figure 4.5 – Configuring the Items parameter in the Filter activity

25. Click **Finish** to continue. Observe that the **Items** parameter is populated. The values from the activities follow the nomenclature `@activity('activity name').output.fieldname`.

26. To configure the filter condition, select the **Condition** text and then click on the **Add dynamic content** link:

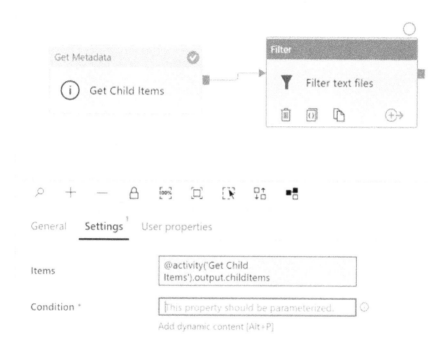

Figure 4.6 – Configuring the Condition parameter in the Filter activity

27. Expand **String Functions** and select **endswith**. Observe that the @endswith function is displayed in the top textbox. In the textbox, enter the following condition, @endswith(item().name,'.txt'). The item().name function returns the name of the file from the Item array as specified in the **Items** parameter of the filter criteria. The endswith function applies to each item from the Item array (the childitems array from the **Get Metadata** activity) and returns true if the filename ends with .txt and false if not:

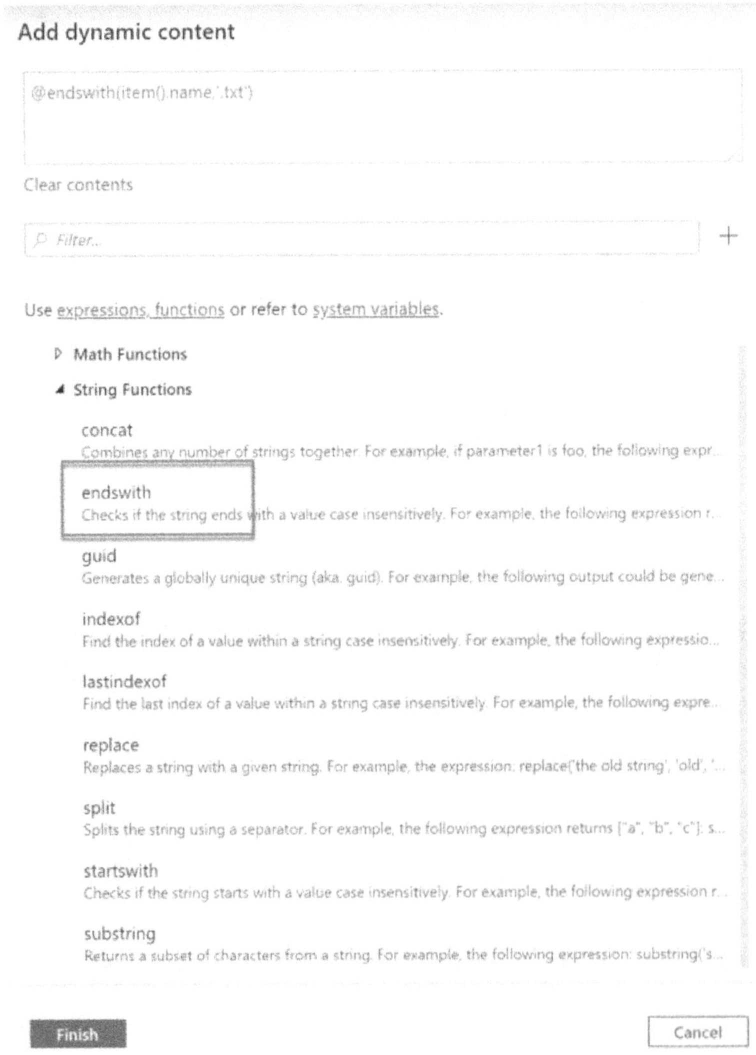

Add dynamic content

@endswith(item().name,'.txt')

Clear contents

🔎 Filter..   +

Use expressions, functions or refer to system variables.

▷ Math Functions

◢ String Functions

concat
Combines any number of strings together. For example, if parameter1 is foo, the following expr...

endswith
Checks if the string ends with a value case insensitively. For example, the following expression r...

guid
Generates a globally unique string (aka. guid). For example, the following output could be gene...

indexof
Find the index of a value within a string case insensitively. For example, the following expressio...

lastindexof
Find the last index of a value within a string case insensitively. For example, the following expre...

replace
Replaces a string with a given string. For example, the expression: replace('the old string', 'old', '...

split
Splits the string using a separator. For example, the following expression returns ['a', 'b', 'c']: s...

startswith
Checks if the string starts with a value case insensitively. For example, the following expression r...

substring
Returns a subset of characters from a string. For example, the following expression: substring('s...

Finish   Cancel

Figure 4.7 – Adding a dynamic filtering condition

28. Click **Finish** to continue. The **Filter** activity should now be configured as shown in the following screenshot:

Figure 4.8 – Configuring the Filter activity

29. Before we proceed further, click **Publish All** to save the changes. Once the changes are published, click on **Debug** to run the pipeline and verify the activity outputs. When the pipeline completes, you should get the following output for the **Get Metadata** activity:

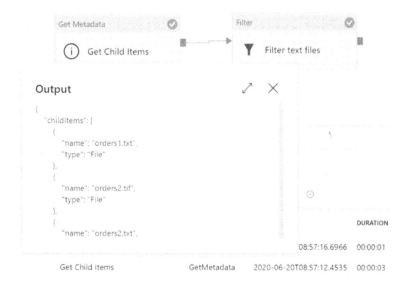

Figure 4.9 – Viewing the Get Metadata activity output

Observe that the output has a `childItems` array with `name` and `type` elements. The output contains all the files from the `orders` dataset. You should get the following output from the **Filter** activity:

Figure 4.10 – Viewing the Filter activity output

Observe that the **Filter** activity output shows `ItemsCount` as 7 and `FilteredItemsCount` as 5. It has filtered the two `.tif` files. The **Filter** activity outputs an array (`Value`) with `name` and `type` elements.

The next step is to iterate and copy each of the files to the `orderstxt` directory.

30. To copy the files, we need to iterate through each of the items in the **Filter** activity and use the **Copy data** activity to copy the file from `orders` to the `orderstxt` dataset. Drag and drop the **ForEach** activity from the **Iteration & conditionals** tab. Connect the **Filter** activity to the **ForEach** activity and then select the **ForEach** activity. Under the **General** tab, rename the **ForEach** activity `For Each File`. Then, switch to the **Settings** tab. The **ForEach** activity requires an array to iterate too. In the **Items** parameter, copy and paste the value `@activity.('Filter text files').output.Value`. The `Value` array is the array returned from the **Filter** activity, as shown in the following screenshot:

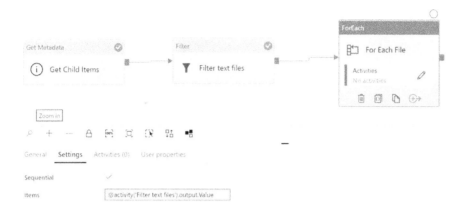

Figure 4.11 – Configuring the ForEach activity

31. Click the pencil icon in the **Activities** box inside the **ForEach** activity. It'll open the
ForEach activity canvas. Drag and drop the **Copy data** activity from the **Move &
transform** tab. Under the **General** tab, rename the **Copy data** activity `Copy file
to orderstxt directory`. Switch to the **Source** tab. Select `ordersfile` as
the source dataset:

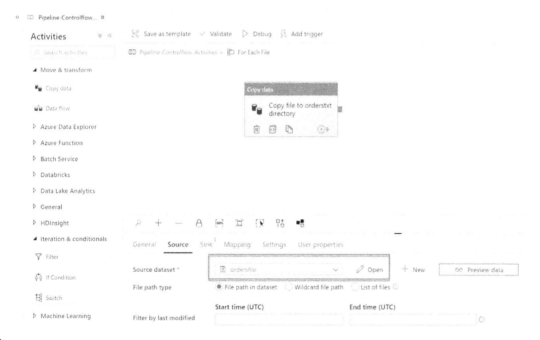

Figure 4.12 – Configuring the Copy data source

The `ordersfile` dataset points to `orders1.txt`. However, we need it to be
dynamic to iterate and copy all the files.

32. To do this, click the **Open** link beside the source dataset:

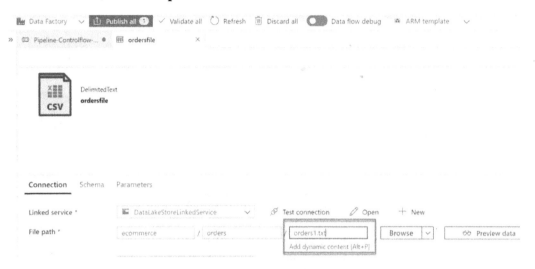

Figure 4.13 – Modifying the ordersfile dataset

33. Click on the `orders1.txt` textbox and then select the **Add dynamic content** link:

Figure 4.14 – Adding dynamic content

34. In the **Add dynamic content** window, click on the plus icon beside the **Filter** textbox. In the **New parameter** window, provide a name of `filename` and a default value of `orders1.txt`:

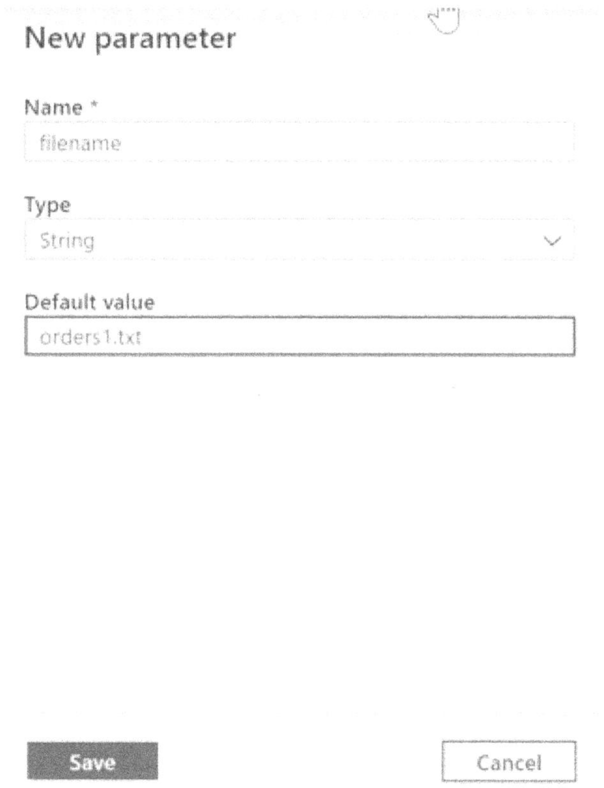

Figure 4.15 – Creating a new parameter

Click **Save** to create the parameter. We are taken back to the **Add dynamic content** window. In the **Add dynamic content** window, scroll down and select the **filename** parameter from the **Parameters** section:

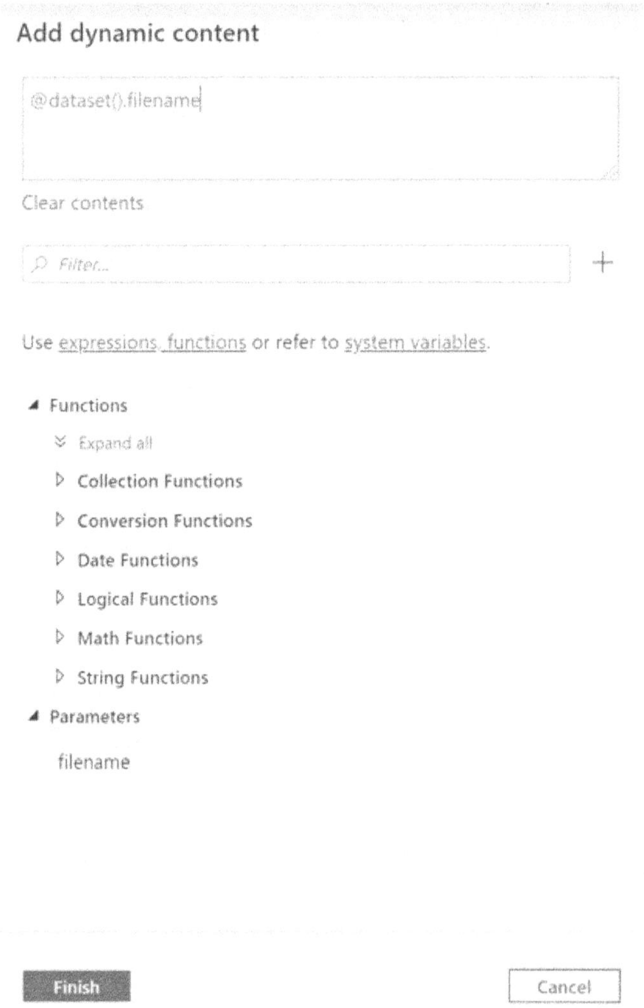

Add dynamic content

> @dataset().filename

Clear contents

> 🔎 Filter...   +

Use expressions, functions or refer to system variables.

▲ Functions

   ⊻ Expand all

   ▷ Collection Functions

   ▷ Conversion Functions

   ▷ Date Functions

   ▷ Logical Functions

   ▷ Math Functions

   ▷ String Functions

▲ Parameters

   filename

Finish   Cancel

Figure 4.16 – Adding dynamic content

35. Click **Finish** to continue.

We are taken back to the `ordersfile` dataset window. Observe that the filename is now dynamic as is specified by the `filename` parameter:

Figure 4.17 – ordersfile dataset window

Note

To preview the data, click on the **Preview data** button. When asked, provide the value for the **filename** parameter as `orders1.txt`.

36. The changes will automatically be reflected in the **Copy** data. Switch to the **Pipeline | ForEach** activity and select the **Copy data** activity, and then the **Source** tab. We need to configure the **filename** parameter to dynamically receive the value from the **ForEach** activity:

Figure 4.18 – Configuring the filename parameter

37. In the **Dataset properties** section, copy and paste @item().name. @item() refers to the item array and name is the name element of the Value array from the **Filter** activity.

38. In the **ForEach | Copy data** activity, select the **Sink** tab. Select the sink dataset as orderstxt. Leave the rest of the values as the default:

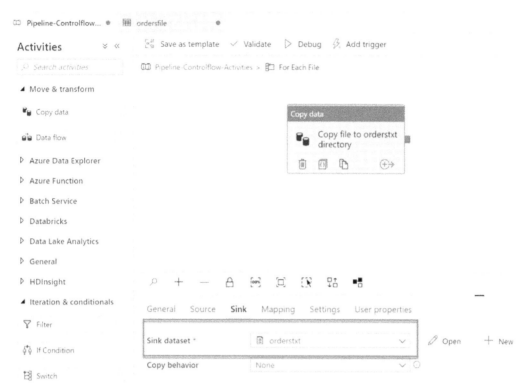

Figure 4.19 – Configuring the Copy data sink dataset

This completes the **Copy data** activity and the pipeline configuration. Click on **Publish all** to save the changes.

The next step is to log the file moved in the database using the stored procedure. To do this, in the **ForEach** loop activity canvas, drag and drop the **Stored Procedure** activity from the **General** tab. Connect the **Copy data** activity to the stored procedure activity. (Note: Before proceeding further, connect to the Azure SQL database and create the `FileCopyStatus` table and the `usp_UpdateCopyStatus` stored procedure as specified in the `~/azure-data-engineering-cookbook\Chapter04\ SQL\FileCopyStatus.sql` file.) In the **General** tab, rename the stored procedure activity to `Log file copy to database`. Switch to the **Settings** tab and select **AzureSqlDatabaseLinkedService** for **Linked Services**. For **Stored procedure name**, select **usp_UpdateCopyStatus** and click **Import parameter**. In the **Stored procedure parameters** section, provide the parameter values as follows:

a) **filename = @item().name**

b) **fromdirectory = orders**

c) **todirectory = orderstxt**

The following is a screenshot for reference:

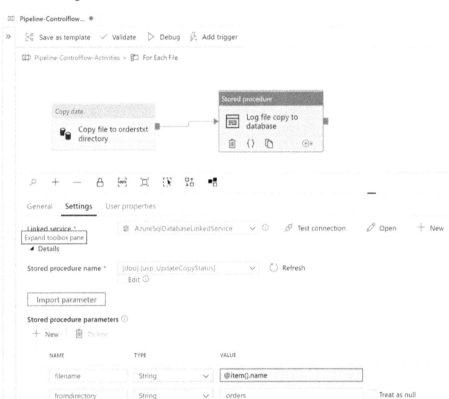

Figure 4.20 – Configuring a stored procedure activity

39. Click **Pipeline-Controlflow-Activities** to return to the pipeline canvas from the **ForEach** canvas:

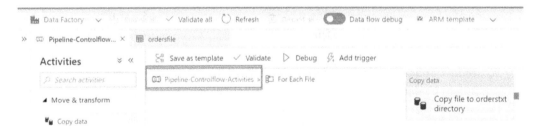

Figure 4.21 – Switching to the pipeline from ForEach

Click **Publish all** to save the pipeline.

40. Before running the pipeline, execute the following command to verify the files in the `orderstxt` directory. There shouldn't be any files as of now:

```
$dls = Get-AzStorageAccount -ResourceGroupName packtade
-Name packtdatalakestore

Get-AzDataLakeGen2ChildItem -FileSystem ecommerce -Path /
orderstxt/ -Context $dls.Context
```

41. Click **Debug** to run the pipeline. Once the pipeline completes, you should get the following output:

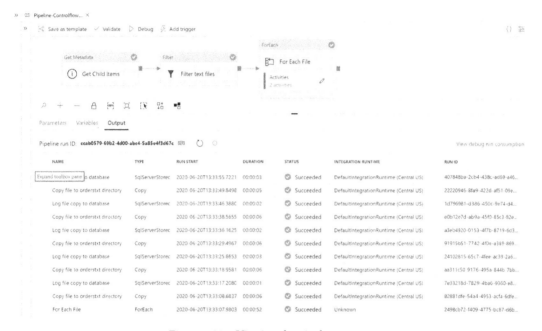

Figure 4.22 – Viewing the pipeline output

42. Execute the following PowerShell command to check the files in the `ordertxt` directory:

```
$dls = Get-AzStorageAccount -ResourceGroupName packtade
-Name packtdatalakestore

Get-AzDataLakeGen2ChildItem -FileSystem ecommerce -Path /
orderstxt/ -Context $dls.Context
```

You should get an output as shown in the following screenshot:

```
PS C:\> Get-AzDataLakeGen2ChildItem -FileSystem ecommerce -Path /orderstxt/ -Context $dls.Context


   FileSystem Name: ecommerce

Path                 IsDirectory Length      LastModified          Permissions Owner        Group
----                 ----------- ------      ------------          ----------- -----        -----
orderstxt/orders1.t… False       14928590    2020-06-20 10:38:42Z  rw-r-----   $superuser   $superuser
orderstxt/orders2.t… False       15082989    2020-06-20 10:38:51Z  rw-r-----   $superuser   $superuser
orderstxt/orders3.t… False       15113409    2020-06-20 10:38:58Z  rw-r-----   $superuser   $superuser
orderstxt/orders4.t… False       15258048    2020-06-20 10:39:06Z  rw-r-----   $superuser   $superuser
orderstxt/orders5.t… False       1915274     2020-06-20 10:39:14Z  rw-r-----   $superuser   $superuser
```

Figure 4.23 – Listing items in the orderstxt directory

Observe that the files with the .txt extension are copied from orders to the orderstxt directory.

43. Connect to the Azure SQL database and execute the following query:

```
SELECT * FROM FileStatusCopy
```

You should get an output as shown in the following screenshot:

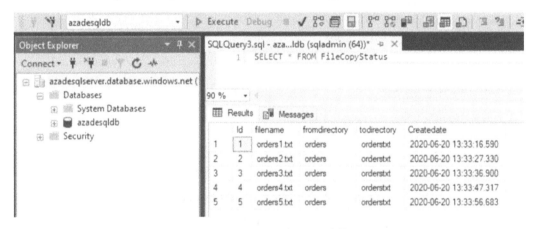

Figure 4.24 – Listing the copied files

Implementing control flow activities – Lookup and If activities

In this recipe, we'll be implementing **Lookup** and **If** control flow activities. The **Lookup** activity, as the name suggests, is used to get a value or an array of values from a given dataset. The If activity, as the name suggests, is used to run tasks based on whether a given condition is true or false.

Getting ready

Before you start, execute the following steps:

1. Log into Azure from PowerShell. To do this, execute the following command and follow the instructions to log into Azure:

   ```
   Connect-AzAccount
   ```

2. Open https://portal.azure.com and log in using your Azure credentials.

3. Create the pipeline as specified in the previous recipe, if not already created.

How to do it...

In the previous recipe, we created a pipeline to copy files from one data lake store directory to another. In this recipe, we'll create a new pipeline to execute the previous pipeline (`Pipeline-Controlflow-Activities`) when the configuration value in the database is true. To create the new pipeline, follow the given steps:

1. In Azure Data Factory, under **Factory Resources**, select the plus icon beside the search box. Select **Pipelines** from the context menu:

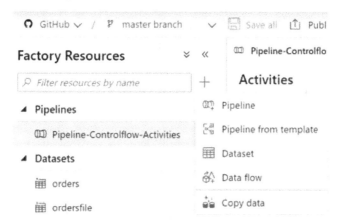

Figure 4.25 – Creating a new pipeline

Name the new pipeline `Pipeline-Controlflow-Activities-2`.

2. In Azure Data Factory, under **Factory Resources**, select the plus icon beside the search box. Select **Datasets** from the context menu. In the **New dataset** window, type sql database in the search box and select **Azure SQL Database**:

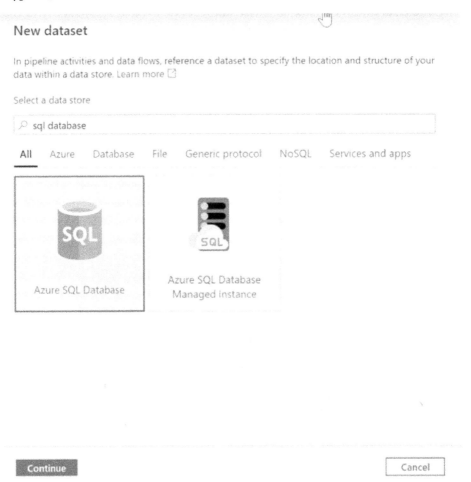

Figure 4.26 – Selecting a data store for a dataset

Click **Continue** to move to the next step.

3. In the **Set properties** window, fill in **Name** with `GetPipelineConfiguration`, **Linked service** with `AzureSqlDatabaseLinkedService`, and leave the rest of the values as the default:

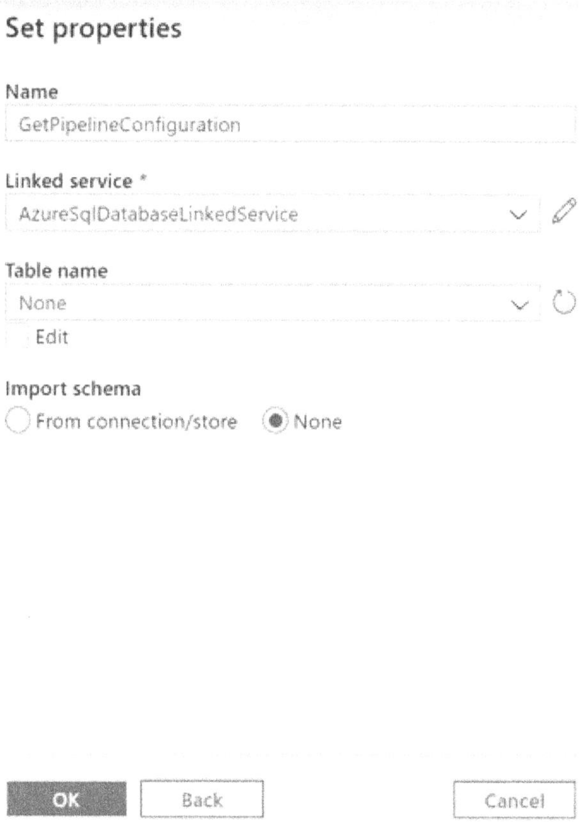

Set properties

Name

GetPipelineConfiguration

Linked service *

AzureSqlDatabaseLinkedService

Table name

None

Edit

Import schema

○ From connection/store   ● None

OK   Back   Cancel

Figure 4.27 – Setting the dataset properties

Click **OK** to create the dataset. The dataset will be used in the lookup activity to get the status of the pipeline.

4. Select the pipeline created in *step 1*. In the pipeline canvas, drag and drop the **Lookup** activity from the **General** tab. In the **Lookup** activity's **General** tab, rename the activity `Get Pipeline Status`. In the **Settings** tab, se **Source dataset** as **GetPipelineConfiguration** and **Use query** as **Stored procedure**. Select the **usp_GetPipelineStatus** stored procedure from the **Name** dropdown. Click **Import parameter**. The **key** parameter will be listed under the **Parameter** heading. Provide the parameter value as `Pipeline-Controlflow-Activities`. Leave the rest of the configuration as the default:

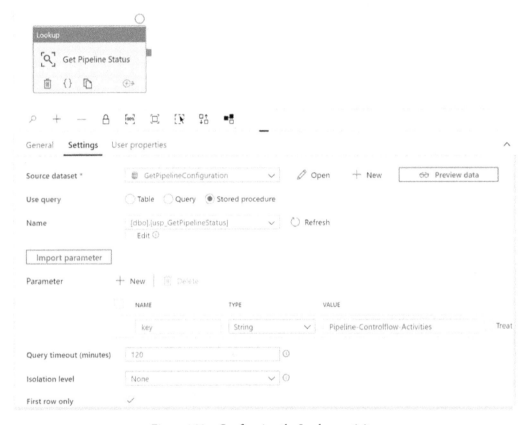

Figure 4.28 – Configuring the Lookup activity

5. Click **Preview data** to verify that we get correct data from the stored procedure. You should get the following in the **Preview data** window:

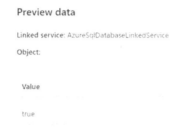

Figure 4.29 – Previewing the Lookup activity data

6. Drag and drop the **If Condition** activity from the **Iteration & conditionals**
    section. Under the **If Condition** activity's **General** tab, rename the activity
    ExecutePipeline. Connect the **Lookup** activity to the **If Condition** activity. In
    the **If Condition** activity's **Activities** tab, copy and paste the following value in the
    Expression textbox:

```
@bool(activity('Get Pipeline Status').output.firstRow.
Value)
```

The preceding code gets the value from the lookup activity and converts it into a
Boolean data type. The **If Condition** activity should be as shown in the following
screenshot:

Figure 4.30 – Configuring the If Condition activity expression

Based on the expression value (**True/False**), we can configure a different set of activities to run. However, for this scenario, we'll only configure activities to run when the expression is true. To do that, click the pen icon in the **ACTIVITY** column beside the **True** value.

7. In the **If Condition True activities** canvas, drag and drop the **Execute Pipeline** activity from the **General** section. The **Execute Pipeline** activity is used to invoke another data factory pipeline from another pipeline. In the **Execution Pipeline General** tab, rename the activity `Execute Pipeline-Controlflow-Activities`. In the **Settings** tab, select **Pipeline-Controlflow-Activities** from the **Invoked pipeline** dropdown. Leave the rest of the configuration values as the default. The **Execute Pipeline** activity should be as shown in the following screenshot:

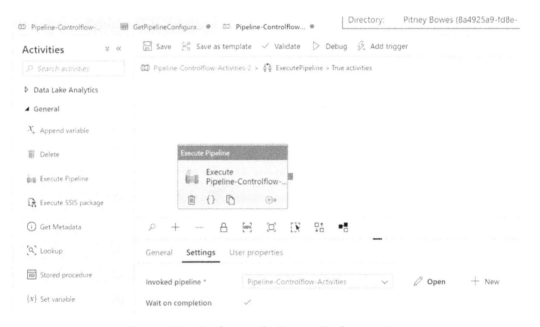

Figure 4.31 – Configuring the Execute Pipeline activity

8. Return to the **Pipeline-Controlflow-Activities2** canvas. Click **Publish all** to save the changes. Click **Debug** to run the pipeline. When the pipeline finishes execution, you should get the following output:

Figure 4.32 – Output after debugging the pipeline

Triggering a pipeline in Azure Data Factory

An Azure Data Factory pipeline can be triggered manually, scheduled, or triggered by an event. In this recipe, we'll configure an event-based trigger to run the pipeline created in the previous recipe whenever a new file is uploaded to the data lake store.

Getting ready

Before you start, perform the following steps:

1. Log into Azure from PowerShell. To do this, execute the following command and follow the instructions to log into Azure:

   ```
   Connect-AzAccount
   ```

2. Open https://portal.azure.com and log in using your Azure credentials.

3. Create the pipeline as specified in the previous recipe, if not already created.

How to do it...

To create the trigger, perform the following steps:

1. The event trigger requires the eventgrid resource to be registered in the subscription. To do that, execute the following PowerShell command:

   ```
   Register-AzResourceProvider -ProviderNamespace Microsoft.
   EventHub
   ```

2. In the Azure portal, under **All resources**, open **packtdatafactory**. On the
    packtdatafactory overview page, select **Author & Monitor**. On the **Data Factory**
    page, select **Author**:

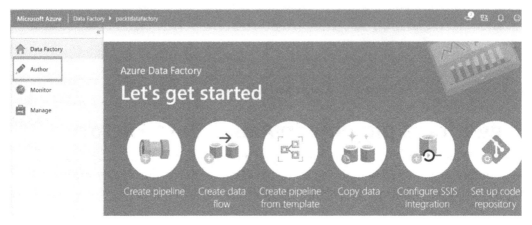

Figure 4.33 – Opening an existing data factory pipeline

3. Under the **Factory Resources** tab, expand the **Pipelines** section and select **Pipeline-
    Controlflow-Activities**:

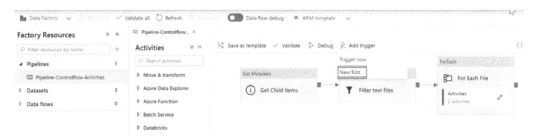

Figure 4.34 – Adding a trigger to a pipeline

Select **Add trigger** and then select **New/Edit**.

> **Note**
> To create **Pipeline-Controlflow-Activities**, refer to the previous recipe.

4. In the **Add triggers** window, select **Choose trigger** and then select **New**.

5. In the **New trigger** window, fill in **Name** as NewFileTrigger. Se **Type** as **Event**. Se **Storage account name** as **packtdatalakestore**, **Container name** as **ecommerce**, and **Blob path begins with** as **orders/**. Se **Event** as **Blob created**. Leave the rest of the settings as the default:

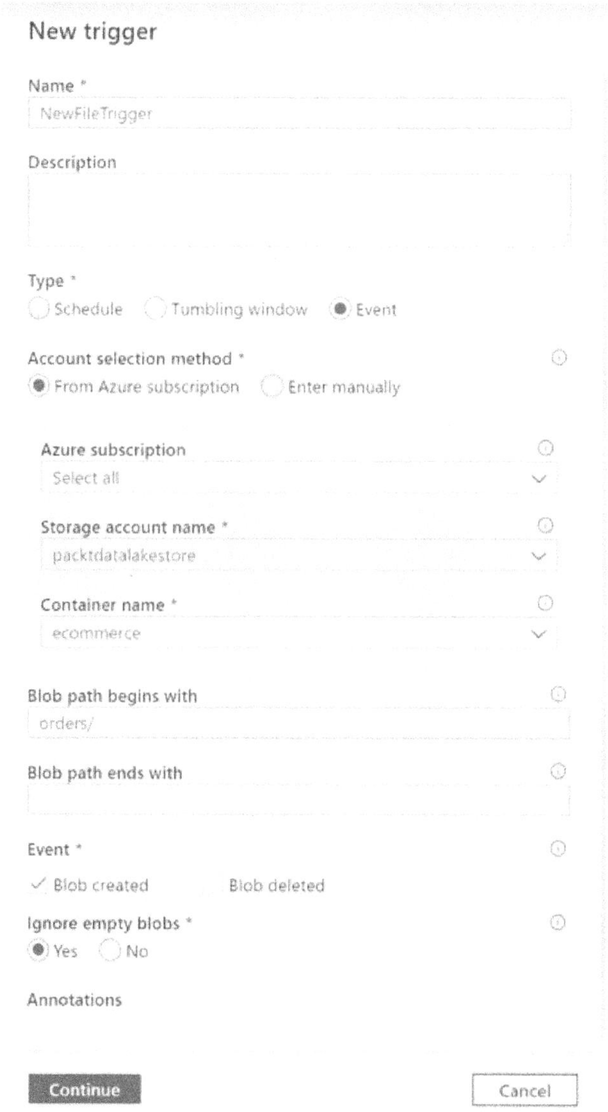

Figure 4.35 – Creating a trigger

Click **Continue** to create the trigger. The trigger will be fired whenever a new file is uploaded to the /ecommerce/orders directory.

6. In the **Data preview** window, all of the files in the `ecommerce/orders` directory will be listed:

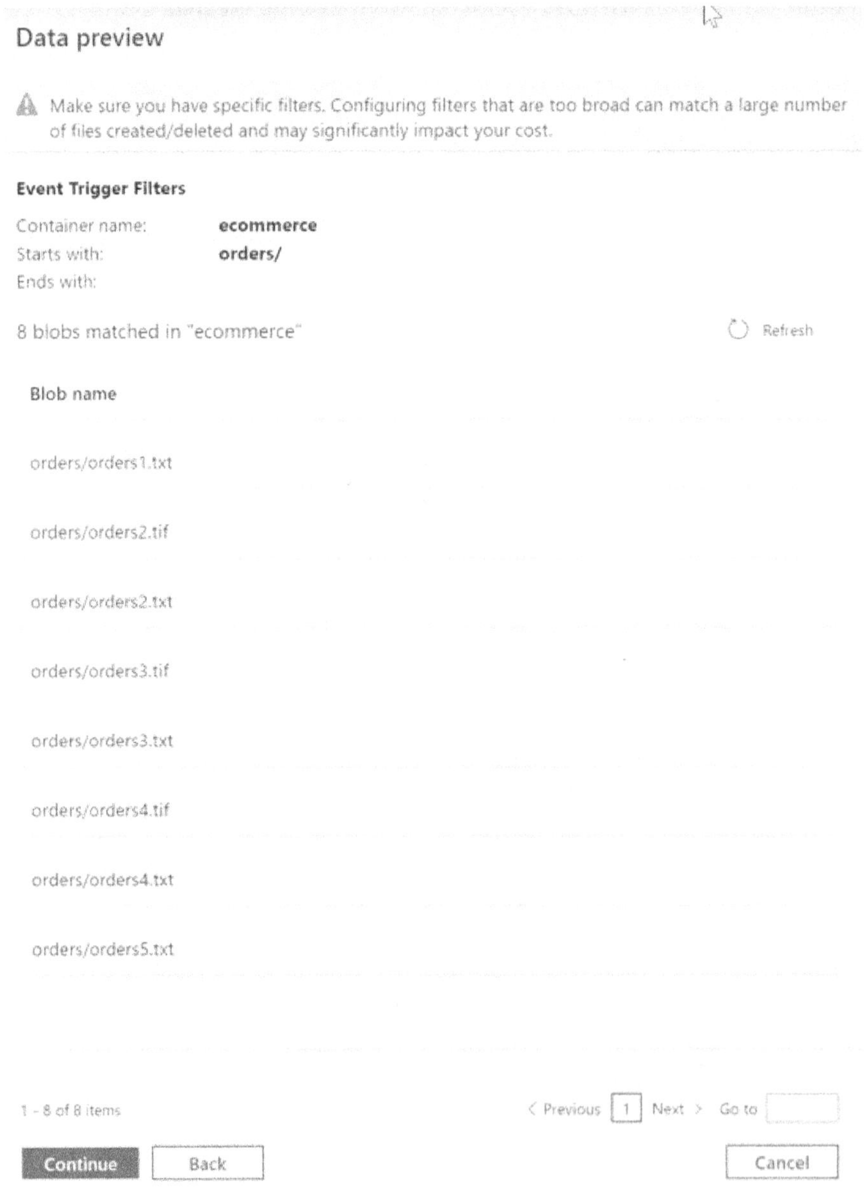

Data preview

⚠ Make sure you have specific filters. Configuring filters that are too broad can match a large number of files created/deleted and may significantly impact your cost.

Event Trigger Filters

Container name: **ecommerce**
Starts with: **orders/**
Ends with:

8 blobs matched in "ecommerce"   ↻ Refresh

Blob name

orders/orders1.txt

orders/orders2.tif

orders/orders2.txt

orders/orders3.tif

orders/orders3.txt

orders/orders4.tif

orders/orders4.txt

orders/orders5.txt

1 - 8 of 8 items   ‹ Previous [1] Next › Go to [   ]

[Continue]   [Back]   [Cancel]

Figure 4.36 – New trigger data preview

Click **Continue** to move to the next step.

7. In the next window, we can specify the parameter values, if any, required by the pipeline to run:

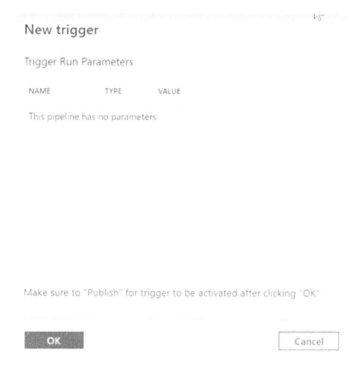

Figure 4.37 – Creating a new trigger

Click **OK** to create the trigger.

8. The trigger is created. Click **Publish all** to save and apply the changes.

9. To see the trigger in action, execute the following command to upload new files to the ecommerce/orders directory:

```
.\ADE\azure-data-engineering-cookbook\Chapter04\6_
UploadData2FoldertoDataLakeStore.ps1 -resourcegroupname
packtade -storageaccountname packtdatalakestore -location
centralus -directory C:\ade\azure-data-engineering-
cookbook\Chapter04\Data2\
```

As and when the files are uploaded, NewFileTrigger will trigger the **Pipeline-Controlflow-Activities** activity and the files will be copied to the orderstxt directory.

10. To check the trigger and pipeline execution, open the **Monitor** window:

Figure 4.38 – Viewing the trigger status

11. Execute the following PowerShell command to verify that the newly uploaded files are copied to the `orderstxt` directory from the `orders` directory:

```
$dls = Get-AzStorageAccount -ResourceGroupName packtade
-Name packtdatalakestore
```

```
Get-AzDataLakeGen2ChildItem -FileSystem ecommerce -Path /
orderstxt/ -Context $dls.Context
```

Observe that the two new files, `order10` and `orders11`, are copied to the `orderstxt` directory:

```
PS C:\> $dls = Get-AzStorageAccount -ResourceGroupName packtade -Name packtdatalakestore
PS C:\>
PS C:\> Get-AzDataLakeGen2ChildItem -FileSystem ecommerce -Path /orderstxt/ -Context $dls.Context


   FileSystem Name: ecommerce

Path                 IsDirectory  Length    LastModified             Permissions  Owner         Group
----                 -----------  ------    ------------             -----------  -----         -----
orderstxt/orders1.t… False        14928590  2020-06-20 14:30:34Z rw-r-----    $superuser    $superuser
orderstxt/orders10.… False        14928590  2020-06-20 14:30:49Z rw-r-----    $superuser    $superuser
orderstxt/orders11.… False        14928590  2020-06-20 14:31:01Z rw-r-----    $superuser    $superuser
orderstxt/orders2.t… False        15082989  2020-06-20 14:31:13Z rw-r-----    $superuser    $superuser
orderstxt/orders3.t… False        15113409  2020-06-20 14:31:27Z rw-r-----    $superuser    $superuser
orderstxt/orders4.t… False        15258048  2020-06-20 14:31:39Z rw-r-----    $superuser    $superuser
orderstxt/orders5.t… False        1915274   2020-06-20 14:31:48Z rw-r-----    $superuser    $superuser
```

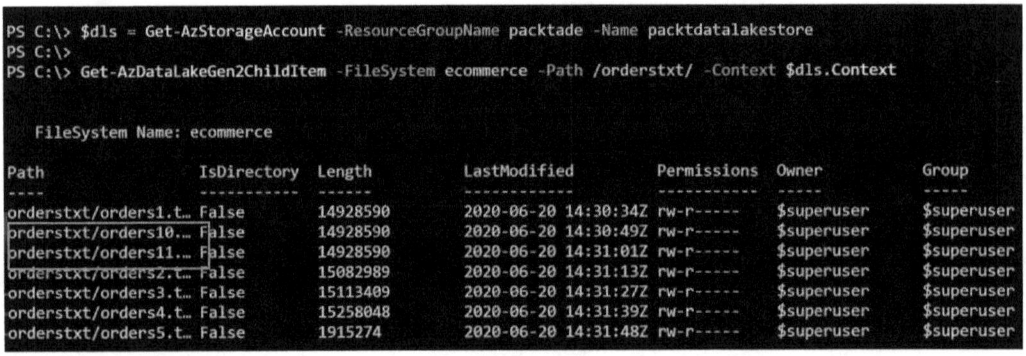

Figure 4.39 – Verifying the pipeline output

5

Control Flow Transformation and the Copy Data Activity in Azure Data Factory

In this chapter, we'll look at the transformation activities available in Azure Data Factory control flows. Transformation activities allow us to perform data transformation within the pipeline before loading data at the source.

In this chapter, we'll cover the following recipes:

- Implementing HDInsight Hive and Pig activities
- Implementing an Azure Functions activity
- Implementing a Data Lake Analytics U-SQL activity

- Copying data from Azure Data Lake Gen2 to an Azure Synapse SQL pool using the copy activity

- Copying data from Azure Data Lake Gen2 to Azure Cosmos DB using the copy activity

Technical requirements

For this chapter, the following are required:

- A Microsoft Azure subscription

- PowerShell 7

- Microsoft Azure PowerShell

Implementing HDInsight Hive and Pig activities

Azure HDInsight is an **Infrastructure as a Service (IaaS)** offering that lets you create big data clusters to use Apache Hadoop, Spark, and Kafka to process big data. We can also scale up or down the clusters as and when required.

Apache Hive, built on top of Apache Hadoop, facilitates querying big data on Hadoop clusters using SQL syntax. Using Hive, we can read files stored in the Apache **Hadoop Distributed File System (HDFS)** as an external table. We can then apply transformations to the table and write the data back to HDFS as files.

Apache Pig, built on top of Apache Hadoop, is a language to perform **Extract, Transform, and Load (ETL)** operations on big data. Using Pig, we can read, transform, and write the data stored in HDFS.

In this recipe, we'll use Azure Data Factory, HDInsight Hive, and Pig activities to read data from Azure Blob storage, aggregate the data, and write it back to Azure Blob storage.

Getting ready

To get started, do the following:

1. Log in to `https://portal.azure.com` using your Azure credentials.

2. Open a new PowerShell prompt. Execute the following command to log in to your Azure account from PowerShell:

```
Connect-AzAccount
```

3. You'll need an existing Data Factory account. If you don't have one, create one by executing the `~/azure-data-engineering-cookbook\Chapter04\3_ CreatingAzureDataFactory.ps1` PowerShell script.

How to do it...

In this recipe, we'll read the sales data stored in Azure Storage to do the following:

- Run a Hive query to calculate sales by country.
- Run a Pig script to calculate quantity by country.

Then, we'll write the output back to Azure Storage using Azure HDInsight Hive and Pig activities, respectively.

Follow the given steps to implement the recipe. Let's start with uploading the data to Azure Storage:

1. Execute the following PowerShell command to create an Azure Storage account, create an `orders` container, and upload the files to the `data` directory in the `orders` container. The script also creates the `ordersummary` container:

```
.\azure-data-engineering-cookbook\Chapter05\1_
UploadDatatoAzureStorage.ps1 -resourcegroupname packtade
-storageaccountname packtstorage -location centralus
-datadirectory .\azure-data-engineering-cookbook\
Chapter05\Data\ -createstorageaccount $true -uploadfiles
$true
```

In the preceding command, if you already have an existing Azure Storage account, then set the `createstorageaccount` parameter to `$false`. You should get an output as shown in the following screenshot:

```
KeyName      : key1
Value        : HUflup+hJXXuK7Qn/n047RMJtUV2U79+ISrNaeb7DhJiF6p+Lv00051nNodaTNBsqp8fKf9KXfoD3mdEJ+NGPg==
Permissions  : Full

KeyName      : key2
Value        : VLXN8OpU/zG90rQmvwW/aBpNeEIvz5DyWx+iPhqh2ZMMwVeaYcjX6Qv9x1q6eKk6n5g/JWimiWFd7qRgm5fcyA==
Permissions  : Full
```

Figure 5.1 – Creating and uploading data to Azure Blob storage

Note down the key value from the preceding screenshot. It'll be different in your case and is required in later steps.

> **Note**
>
> The command used in the preceding PowerShell script is explained in *Chapter 1, Working with Azure Blob Storage.*

2. The next step is to create an Azure Data Factory linked service for the storage account we created in the previous step. To do this, open `~/Chapter05/PacktStorageLinkedServiceDefinition.txt` in a notepad. For `connectionString`, provide the storage account name and account key for the storage account created in the previous step. `connectionString` should be as shown:

```
"connectionString": "DefaultEndpointsProtocol=https;
AccountName=packtstorage;EndpointSuffix=core.
windows.net;AccountKey=234f/
vOAZ3jd8TGZ1DGILZqtU+VQcnjHd2D879viyeax0XDTC8k2ji7
YFcg6D1WxBziH38tNwtRmdJOCdq1/8Q==",
```

> **Note**
>
> The storage account name and key may be different in your case.

3. Save and close the file. Execute the following PowerShell command to create the linked service:

```
Set-AzDataFactoryV2LinkedService -Name
PacktStorageLinkedService -DefinitionFile C:\
azure-data-engineering-cookbook\Chapter05\
PacktStorageLinkedServiceDefinition.txt
-ResourceGroupName packtade -DataFactoryName
packtdatafactory
```

> **Note**
>
> You may have to change the parameter values as per your environment.

4. The next step is to create an HDInsight linked service. We can either create a linked service to connect to an existing HDInsight cluster or we can spin up an on-demand HDInsight cluster. We'll use an on-demand HDInsight cluster. To provision an on-demand HDInsight cluster, Azure Data Factory requires a service principal ID and password. Execute the following PowerShell command to create the service principal and grant it permission to provision Azure resources in the `packtade` resource group:

```
.\azure-data-engineering-cookbook\Chapter05\2_
CreateServicePrincipal.ps1 -resourcegroupname packtade
-password "Awesome@1234"
```

The preceding command runs the `New-AzADServicePrincipal` cmdlet to create a new service principal and the `New-AzADAppCredential` cmdlet to create the service application credentials.

You should get an output as shown in the following screenshot:

Figure 5.2 – Creating a service principal

Note down the application ID of the service principal as it's required in the next step.

5. To create a linked service for the HDInsight cluster, open the `~/Chapter05/HDInsightHadoopLinkedServiceDefinition.txt` definition file. Observe that the linked service type is `HDInsightOnDemand`. The `timeToLive` value is 15 minutes. This means that the cluster will only be available for 15 minutes. We can decrease or increase `timeToLive` as required. For example, if a Hive query takes 30 minutes to run, set `timeToLive` to be greater than 30 minutes. In the definition file, do the following:

a) Change the service principal ID as copied in *step 4*.

b) Change the password as provided in *step 4*.

c) Change the tenant. To get the tenant, execute the (Get-AzSubscription). TenantId PowerShell command.

Save and close the definition file.

Execute the following PowerShell command to create the linked service:

```
Set-AzDataFactoryV2LinkedService -Name
HDInsightHadoopLinkedService -DefinitionFile C:\
azure-data-engineering-cookbook\Chapter05\
HDInsightHadoopLinkedServiceDefinition.txt
-ResourceGroupName packtade -DataFactoryName
packtdatafactory
```

You will have to change the parameter values as per your environment. You should get an output as shown in the following screenshot:

```
PS C:\> Set-AzDataFactoryV2LinkedService -Name HDInsightHiveLinkedService -DefinitionFile C:\azure-data-
engineering-cookbook\Chapter7\HDInsightHiveLinkedServiceDefinition.txt -ResourceGroupName packtade -Data
FactoryName packtdatafactory

LinkedServiceName  : HDInsightHiveLinkedService
ResourceGroupName  : packtade
DataFactoryName    : packtdatafactory
Properties         : Microsoft.Azure.Management.DataFactory.Models.HDInsightOnDemandLinkedService
```

Figure 5.3 – Creating a linked service for an HDInsight on-demand cluster

6. The next step is to create the pipeline. To do this, open Data Factory and create a new pipeline. Name the pipeline Pipeline-Hive-Pig-Activity:

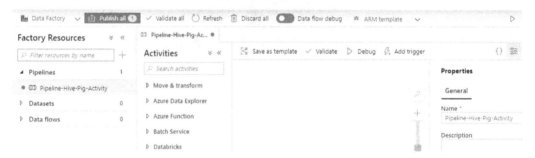

Figure 5.4 – Creating a new pipeline

7. From the **Activities** bar, expand **HDInsight** and drag and drop the Hive and Pig activities onto the pipeline canvas:

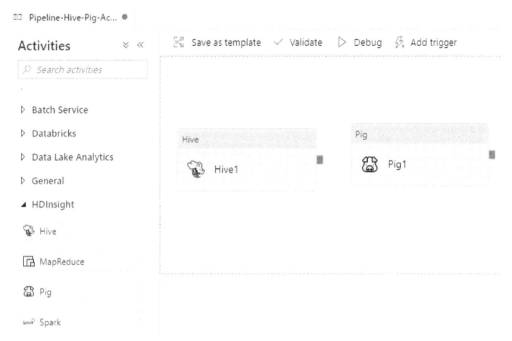

Figure 5.5 – Adding the Hive and Pig activities

8. Select the **Hive** activity and then select the **General** tab. Rename the
    activity to `SalesbyCountry`. Switch to the **HDI Cluster** tab. Select
    HDInsightHadoopLinkedService in the **HDInsight linked service** dropdown:

Figure 5.6 – Selecting the HDInsight linked service

9. Switch to the **Script** tab. Select **PacktStorageLinkedService** in the **Script linked
    service** dropdown. The script linked service storage is where the Hive script can
    be uploaded.

> **Note**
>
> We can either upload a file to the script linked service storage account or upload a local file path. It's always better to upload the script to Azure Storage.

10. For **File path**, select the **Browse local** button and select the `~\Chapter05\ SalesbyCountry.hql` file:

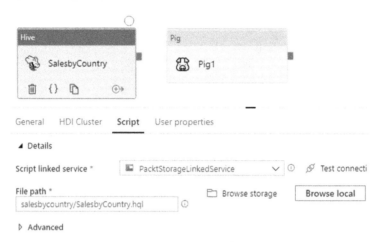

Figure 5.7 – Providing a Hive query

This completes the Hive activity configuration.

11. Select the **Pig** activity and then select the **General** tab. Rename the activity to `QuantitybyCountry`. Switch to the **HDI Cluster** tab and select **HDInsightHadoopLinkedService** for **HDInsight linked service**:

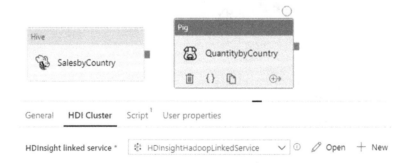

Figure 5.8 – Selecting the HDInsight linked service

12. Switch to the **Script** tab. Select **PacktStorageLinkedService** in the **Script linked service** dropdown. Select **Browse local** and select the `~/Chapter05/ QuantitybyCountry.pig` file:

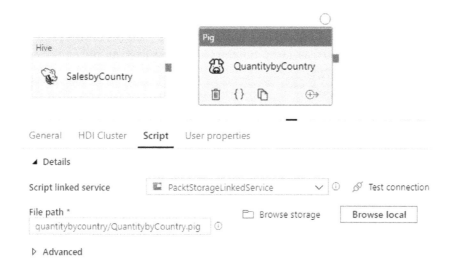

Figure 5.9 – Providing a Pig query

This completes the pipeline configuration. Click **Publish all** to save your work.

13. To run the pipeline, click **Debug**. The pipeline will take around 30–40 minutes to complete. The HDInsight cluster provisioning takes time. You should see the following output when the job completes:

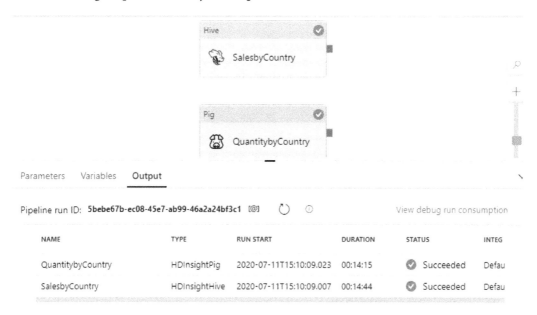

Figure 5.10 – Viewing the pipeline output

14. In the Azure portal, under **All resources**, select the `packtstorage` account. Navigate to **Overview | Containers | ordersummary | salesbycountry**. Observe how a new blob is created as a result of the Hive query execution:

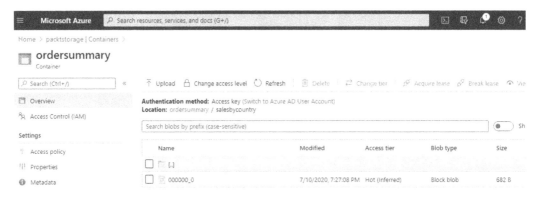

Figure 5.11 – Viewing the HDInisght Hive activity output

15. Again, navigate to **Overview | Containers | ordersummary | quantitybycountry**. Observe how a new blob is created as a result of the Pig script execution:

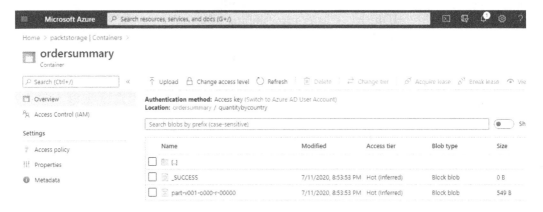

Figure 5.12 – Viewing the Pig activity output

Implementing an Azure Functions activity

Azure Functions is a serverless compute service that lets us run code without the need for any virtual machine or containers. In this recipe, we'll implement an Azure Functions activity to run an Azure function to resume an Azure Synapse SQL database.

Getting ready

To get started, do the following:

1. Log in to `https://portal.azure.com` using your Azure credentials.

2. Open a new PowerShell prompt. Execute the `Connect-AzAccount` command to log in to your Azure account from PowerShell.

3. You will need an existing Data Factory account. If you don't have one, create one by executing the `~/azure-data-engineering-cookbook\Chapter04\3_CreatingAzureDataFactory.ps1` PowerShell script.

How to do it...

Let's start by creating an Azure function to resume an Azure Synapse SQL database:

1. In the Azure portal, type `functions` in the **Search** box and select **Function App** from the search results:

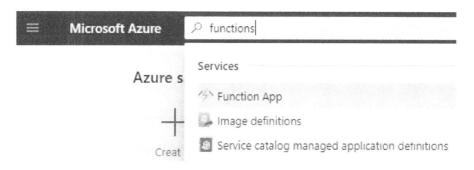

Figure 5.13 – Selecting the Function App service

2. On the **Create Function App** page, provide a resource group name and a function app name. Select **Powershell Core** for **Runtime stack**:

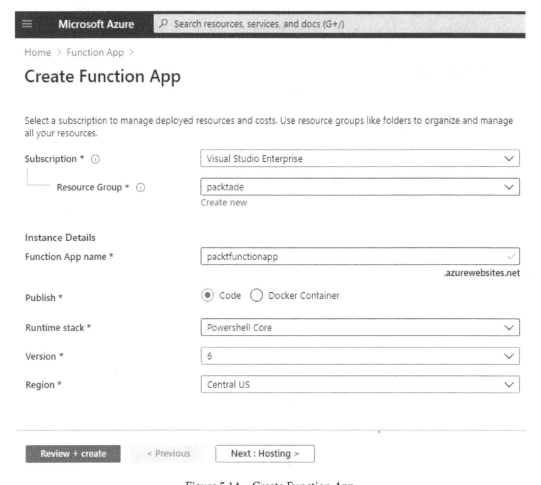

Figure 5.14 – Create Function App

3. Click **Review + create** and then **Create** to create the function app. Wait for the function app to be provisioned.

4. Navigate to the function app created in *step 2*. Find and select **Functions**, and then click **+ Add**:

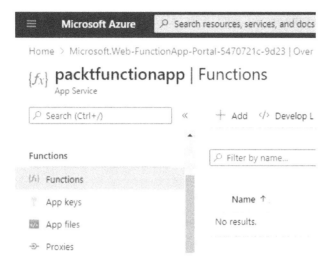

Figure 5.15 – Adding a function to the function app

5. On the **New Function** page, select the **HTTP trigger** template. Name the
    function resumesynapsesql. Leave **Authorization level** as **Function** and
    click **Create Function**:

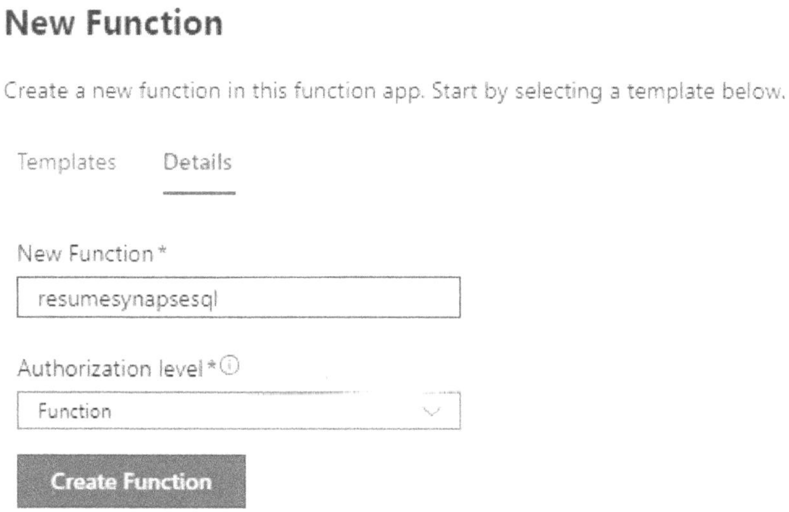

New Function

Create a new function in this function app. Start by selecting a template below.

Templates   Details

New Function *

resumesynapsesql

Authorization level * ⓘ

Function

Create Function

Figure 5.16 – Creating a function

6. On the **resumesynapsesql** page, select **Code + Test**. Copy the code from the ~\Chapter05\ 3_AzureFunction.ps1 file and paste it in the **resumesynapsesql | Code + Test** window. The code takes the tenant ID, service application ID, password, resource group name, server name, and database name as parameters. It uses Connect-AzAccount to connect to the Azure account using the service principal account credentials and the tenant ID and resumes the Synapse SQL database using the Resume-AzSqlDatabase cmdlet:

Figure 5.17 – Adding function code

> **Note**
>
> The Azure function uses the service principal ID for authentication. If you don't have one, follow *step 3* in the *Implementing HDInsight Hive and Pig activities* recipe.

7. The function will resume the Synapse SQL pool and will send the message Synapse Sql Pool Status: online. Click **Save** to save the code. Click **Get Function Url**, then copy the code from the URL and keep it for later use:

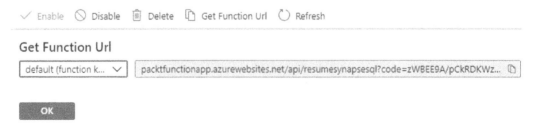

Figure 5.18 – Copying the function key

This completes the Azure function configuration.

8. We'll now provision a new Azure Synapse SQL database. To do that, execute the following PowerShell script:

```
.\azure-data-engineering-cookbook\Chapter05\4_
ProvisioningAzureSynapseSqlPool.ps1 -resourcegroupname
packtade -servername azadesqlserver -databasename
azadesqlpool -sqladmin sqladmin -sqladminpassword "Sql@
Server@1234" -location centralus
```

The preceding script does the following:

a) Creates a resource group if it doesn't exist

b) Creates an Azure SQL Server instance

c) Creates an Azure Synapse SQL database

d) Suspends/pauses the Azure Synapse SQL database

9. The next step is to create a linked service for the Azure function. To do that, open `~\Chapter05\Azurefuntiondefinitionfile.txt`. If you have used a different function app name, replace `functionAppUrl` accordingly. Save and close the file. Execute the following command to create an Azure function linked service:

```
Set-AzDataFactoryV2LinkedService -Name
AzureFunctionLinkedService -DefinitionFile .\
azure-data-engineering-cookbook\Chapter05\
AzureFunctionLinkedService.txt -ResourceGroupName
packtade -DataFactoryName packtdatafactory
```

10. The next step is to create the pipeline. Open Data Factory and create a new pipeline. Name the pipeline `Pipeline-AzureFunction-Activity`. Drag and drop the Azure Functions activity. Under the **General** tab, change the activity name to `Resume SQL Synapse pool`. Under the **Settings** tab, from the **Azure Function linked service** dropdown, select **AzureFunctionLinkedService**. Provide the function name `resumesqlsynapse`. Select **POST** for **Method**. Copy and paste the following code in the **Body** textbox:

```
{
    "name": "Azure",
    "Tenantid":"8a4925a9-fd8e-4866-b31c-f719fb05dce6",
    "user":"60d3d907-566c-44cf-bf7b-77edfce46f12",
    "password":"Awesome@1234",
    "ResourceGroupName":"packtade",
```

```
    "ServerName":"azadesqlserver",
    "DatabaseName":"azadesqlpool"
}
```

> **Note**
>
> The body has the parameter values to be passed to the Azure function. These may differ in your case. The `user` parameter is the service principal ID and `password` is the service principal ID password.

The screen will look as follows:

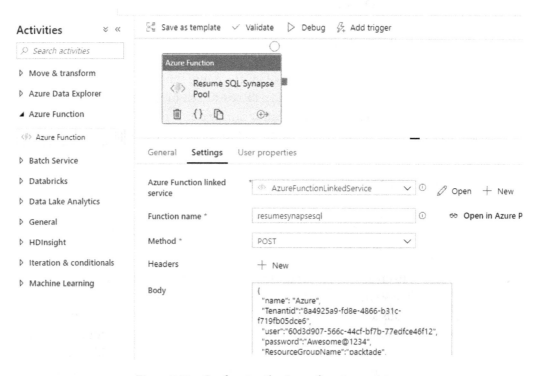

Figure 5.19 – Configuring the Azure function activity

11. Click **Publish all** to save your work, and then click **Debug** to execute the pipeline. You should get the output shown in the following screenshot:

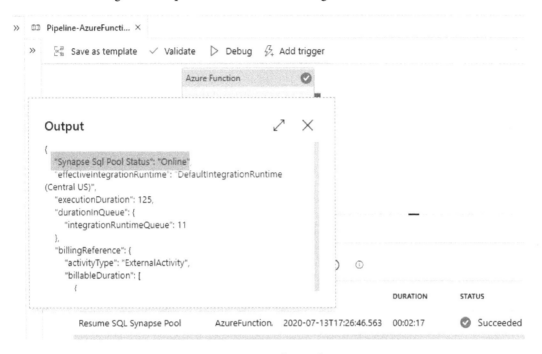

Figure 5.20 – Viewing the pipeline output

> **Note**
>
> You should either delete or pause the Synapse SQL pool to save costs if it's not required anymore.

Implementing a Data Lake Analytics U-SQL activity

Azure Data Lake Analytics is an on-demand analytics service that allows you to process data using R, Python, and U-SQL without provisioning any infrastructure. All we need to do is to upload the data onto Data Lake, provision the Data Lake Analytics account, and run U-SQL to process the data.

In this recipe, we'll implement a Data Lake Analytics U-SQL activity to calculate total sales by country from the `orders` data stored in the Data Lake store.

Getting ready

To get started, do the following:

1. Log in to `https://portal.azure.com` using your Azure credentials.

2. Open a new PowerShell prompt. Execute the `Connect-AzAccount` command to log in to your Azure account from PowerShell.

3. You will need an existing Data Factory account. If you don't have one, create one by executing the `~/azure-data-engineering-cookbook\Chapter04\3_CreatingAzureDataFactory.ps1` PowerShell script.

How to do it...

Let's start by creating a Data Lake Storage Gen1 account and uploading the `orders` file:

1. Execute the following PowerShell command to create a new Data Lake Storage Gen1 account:

```
New-AzDataLakeStoreAccount -ResourceGroupName packtade
-Name packtdatalakestore -Location centralus
```

The preceding command creates a new Data Lake Storage Gen1 account, `packtdatalakestore`, in the `packtade` resource group.

2. Execute the following command to upload the `orders` file to the Data Lake Storage Gen1 account:

```
#Create the orders/data folder
New-AzDataLakeStoreItem -Account packtdatalakestore -Path
/orders/data -Folder
#Upload the orders11.csv file.
New-AzDataLakeStoreItem -Account packtdatalakestore
-Path /orders/data/orders11.csv -Value .\azure-data-
engineering-cookbook\Chapter05\Data\orders11.csv
```

The preceding command creates a new folder called `/orders/data` and then uploads the `orders11.csv` file in it. The `orders11.csv` file has a list of country names along with the product quantity and unit price.

3. Execute the following command to create a new Data Lake Analytics account:

```
New-AzDataLakeAnalyticsAccount -ResourceGroupName
packtade -Name packtdatalakeanalytics -Location centralus
-DefaultDataLakeStore packtdatalakestore
```

The preceding command creates a new Data Lake Analytics account,
`packtdatalakeanalytics`, with default storage as `packtdatalakestore`.

4. We need to have a service principal to access the Data Lake Analytics account from
the data factory. Execute the following command to create a service principal:

```
.\azure-data-engineering-cookbook\Chapter05\2_
CreateServicePrincipal.ps1 -resourcegroupname packade
-password "Awesome@1234"
```

Note down the application ID as it's required in the later steps.

5. We need to grant permission to the service principal created in the previous step to
access the Data Lake Analytics service. To do that, in the Azure portal, navigate to
the Data Lake Analytics account created in *step 1*. Type add user in the **Search**
box and then select **Add user wizard**:

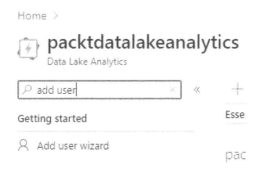

Figure 5.21 – Adding a new user

6. In the **Add user wizard** dialog, click **Select user** and then **Search,** and then select hdinsightsp (the service principal created in *step 4*). Select **Data Lake Analytics Developer** for **Role:**

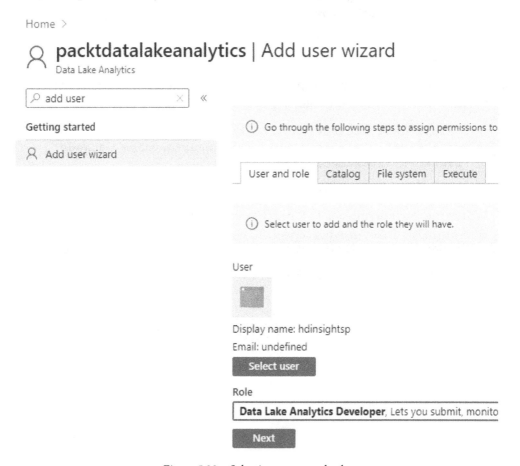

Figure 5.22 – Selecting a user and role

Click **Next** to continue.

7. Under the **File system** tab, the default Data Lake store for the Analytics account is listed. Check the **Read**, **Write,** and **Execute** permissions and for **Apply,** select **This folder and all children** for all the listed paths:

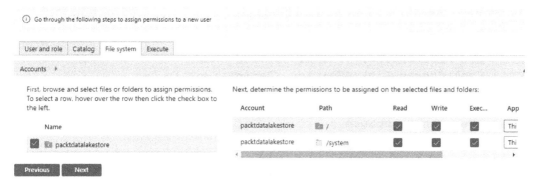

Figure 5.23 – Giving permissions to the Data Lake store

Click **Next**, and then click **Run** to grant the permissions.

8. The next step is to create a storage account, which is required by Data Factory to store/access the U-SQL script. To do that, execute the following PowerShell script:

```
.\azure-data-engineering-cookbook\Chapter05\1_
UploadDatatoAzureStorage.ps1 -resourcegroupname packtade
-storageaccountname packtstorage -location centralus
-uploadfiles $false
```

Note the storage account key as it's required in later steps.

9. The next step is to create an Azure Data Factory linked service for the storage account we created in the previous step. To do this, open ~/Chapter05/ PacktStorageLinkedServiceDefinition.txt in a notepad. For connectionString, provide the storage account name and account key for the storage account created in the previous step. connectionString should be as shown:

```
"connectionString": "DefaultEndpointsProtocol=https;
AccountName=packtstorage;EndpointSuffix=core.windows.net;
AccountKey=234f/vOAZ3jd8TGZ1DGILZqtU+VQcnjHd2D879viyeax0
XDTC8k2ji7YFcg6D1WxBziH38tNwtRmdJOCdq1/8Q==",
```

> **Note**
> The storage account name and key may be different in your case.

10. Save and close the file. Execute the following PowerShell command to create the linked service:

```
Set-AzDataFactoryV2LinkedService -Name
PacktStorageLinkedService -DefinitionFile C:\
azure-data-engineering-cookbook\Chapter05\
PacktStorageLinkedServiceDefinition.txt
-ResourceGroupName packtade -DataFactoryName
packtdatafactory
```

11. We'll now create a linked service for the Data Lake Analytics account. Open `~/Chapter05 Chapter05/DataLakeAnalyticsLinkedService.txt` and provide the following property values: `accountName` (the Data Lake Analytics account name), `servicePrincipalId` (the ID from *step 4*), `servicePrincipalKey.Value` (the service principal ID password), `tenantid`, `subscriptionid`, and `resourceGroupName`. Save and close the file. Execute the following command to create the linked service:

```
Set-AzDataFactoryV2LinkedService -Name
DataLakeAnlayticsLinkedService -DefinitionFile .\
azure-data-engineering-cookbook\Chapter05\
DataLakeAnalyticsLinkedServiceDefinition.
txt -ResourceGroupName packtade -DataFactoryName
packtdatafactory
```

12. The next step is to create the pipeline. Open Data Factory and create a new pipeline. Name the pipeline `Pipeline-USQL-Activity`. Drag and drop the **Data Lake Analytics U-SQL** activity. Under the **General** tab, rename the activity to `SalesbyCountry`. Under the **ADLA Account** tab, select **DataLakeAnalyticsLinkedService** from the **ADLA Linked service** dropdown. Under the **Script** tab, select **PacktStorageLinkedService** from the **Script linked service** dropdown. Select **Browse local**, and then select `~/Chapter05/SalesByCountry.usql`:

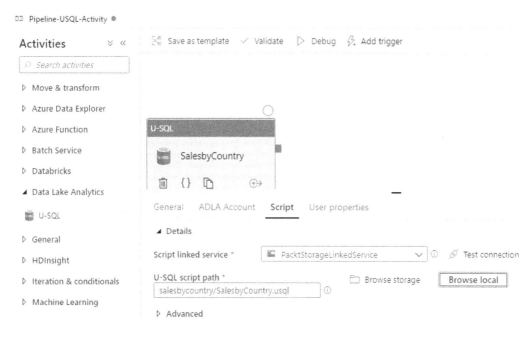

Figure 5.24 – Configuring the U-SQL activity

13. Select **Publish all** to save your work. Click **Debug** to run the pipeline.

14. When the pipeline completes, navigate to the Data Lake Storage account in the Azure portal. A new file, `salesbycountry.csv`, is added to the `/orders/data` folder as a result of the U-SQL query execution from the Data Factory pipeline.

Copying data from Azure Data Lake Gen2 to an Azure Synapse SQL pool using the copy activity

The copy activity, as the name suggests, is used to copy data quickly from a source to a destination. In this recipe, we'll learn how to use the copy activity to copy data from Azure Data Lake Gen2 to an Azure Synapse SQL pool.

Getting ready

Before you start, do the following:

1. Log in to Azure from PowerShell. To do this, execute the `Connect-AzAccount` command and follow the instructions to log in to Azure.

2. Open `https://portal.azure.com` and log in using your Azure credentials.

How to do it...

Follow the given steps to perform the activity:

1. The first step is to create a new Azure Data Lake Gen2 storage account and upload the data. To create the storage account and upload the data, execute the following PowerShell command:

    ```
    .\ADE\azure-data-engineering-cookbook\Chapter04\1_
    UploadOrderstoDataLake.ps1 -resourcegroupname packtade
    -storageaccountname packtdatalakestore -location
    centralus -directory C:\ADE\azure-data-engineering-
    cookbook\Chapter03\Data\
    ```

    > **Note**
    >
    > The PowerShell script is the same as described in the *Loading data in a SQL pool using PolyBase with T-SQL* recipe of *Chapter 5, Analyzing Data with Azure Synapse Analytics*. You may have to change the parameter values as Azure Storage account names are unique globally across Microsoft Azure.

    The command produces several outputs; however, you should get a final output as shown in the following screenshot:

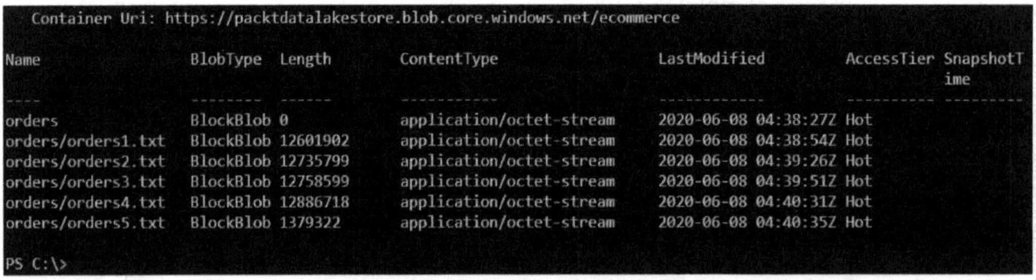

    ```
    Container Uri: https://packtdatalakestore.blob.core.windows.net/ecommerce

    Name                BlobType    Length    ContentType                 LastModified            AccessTier SnapshotT
                                                                                                             ime
    ----                --------    ------    -----------                 ------------            ---------- ---------
    orders              BlockBlob   0         application/octet-stream    2020-06-08 04:38:27Z Hot
    orders/orders1.txt  BlockBlob   12601902  application/octet-stream    2020-06-08 04:38:54Z Hot
    orders/orders2.txt  BlockBlob   12735799  application/octet-stream    2020-06-08 04:39:26Z Hot
    orders/orders3.txt  BlockBlob   12758599  application/octet-stream    2020-06-08 04:39:51Z Hot
    orders/orders4.txt  BlockBlob   12886718  application/octet-stream    2020-06-08 04:40:31Z Hot
    orders/orders5.txt  BlockBlob   1379322   application/octet-stream    2020-06-08 04:40:35Z Hot

    PS C:\>
    ```

    Figure 5.25 – Uploading data to Data Lake Storage

2. We created the source in the previous step, so we'll now create the destination. Execute the following PowerShell command to provision a new Azure Synapse SQL pool:

    ```
    .\ADE\azure-data-engineering-cookbook\Chapter04\2_
    ProvisioningAzureSynapseSqlPool.ps1 -resourcegroupname
    packtade -location centralus -sqlservername
    packtsqlserver -sqlpoolname packtdw -sqladminuser
    sqladmin -sqlpassword Sql@Server@1234
    ```

Note

The PowerShell script is the same as described in the *Provisioning and connecting to an Azure Synapse SQL pool using PowerShell* recipe of *Chapter 5, Analyzing Data with Azure Synapse Analytics*. You may have to change the parameter values as Azure SQL Server names are unique globally across Microsoft Azure.

The script creates and suspends (pauses) the SQL pool to save costs. We'll enable it later when required. You should get a similar output to what's shown in the following screenshot:

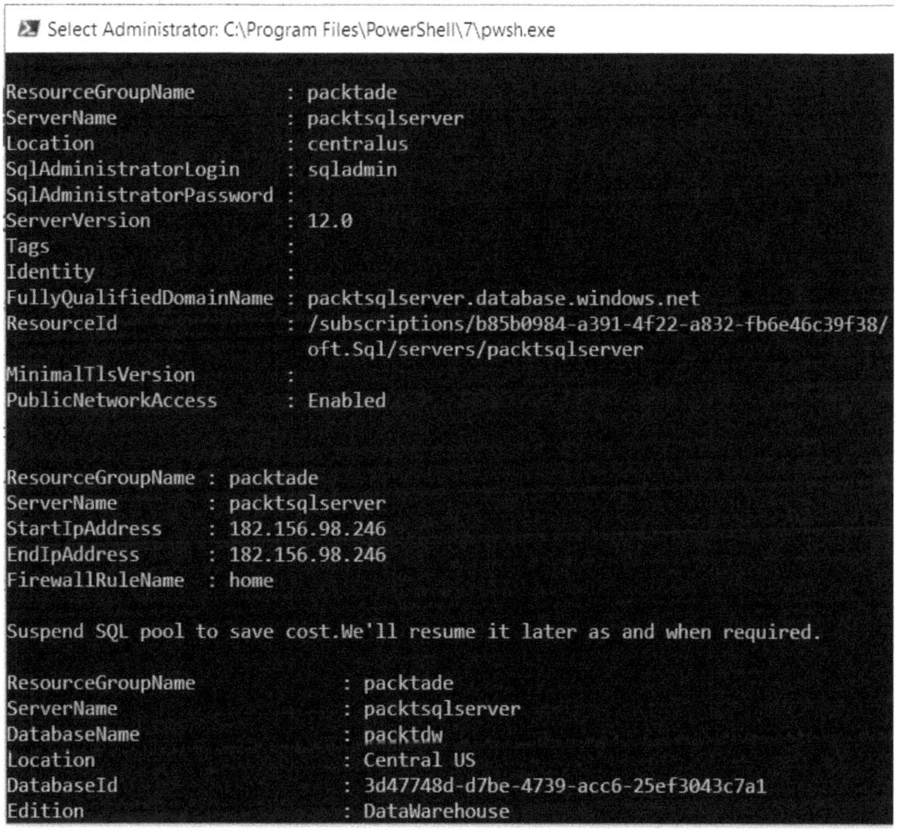

```
Select Administrator: C:\Program Files\PowerShell\7\pwsh.exe

ResourceGroupName        : packtade
ServerName               : packtsqlserver
Location                 : centralus
SqlAdministratorLogin    : sqladmin
SqlAdministratorPassword :
ServerVersion            : 12.0
Tags                     :
Identity                 :
FullyQualifiedDomainName : packtsqlserver.database.windows.net
ResourceId               : /subscriptions/b85b0984-a391-4f22-a832-fb6e46c39f38/
                           oft.Sql/servers/packtsqlserver
MinimalTlsVersion        :
PublicNetworkAccess      : Enabled


ResourceGroupName : packtade
ServerName        : packtsqlserver
StartIpAddress    : 182.156.98.246
EndIpAddress      : 182.156.98.246
FirewallRuleName  : home

Suspend SQL pool to save cost.We'll resume it later as and when required.

ResourceGroupName        : packtade
ServerName               : packtsqlserver
DatabaseName             : packtdw
Location                 : Central US
DatabaseId               : 3d47748d-d7be-4739-acc6-25ef3043c7a1
Edition                  : DataWarehouse
```

Figure 5.26 – Creating a new Azure Synapse SQL pool

3. Execute the following command to create an Azure data factory:

```
.\ADE\azure-data-engineering-cookbook\Chapter04\3_
CreatingAzureDataFactory.ps1 -resourcegroupname packtade
-location centralus -datafactoryname packtdatafactory
```

The script uses the `Set-AzDataFactoryV2` command to create a new Azure Data Factory instance. You should get the following output:

Figure 5.27 – Creating an Azure data factory

4. We now have the source and destination ready. The next step is to create an Azure Data Factory pipeline to use the copy activity to copy the data from the source to the destination. To configure the Azure Data Factory pipeline, we will do the following:

a) Create the Azure Data Factory pipeline.

b) Create the linked service for the source and destination.

c) Configure the copy activity.

5. In the Azure portal, navigate to **All resources** and open `packtdatafactory`:

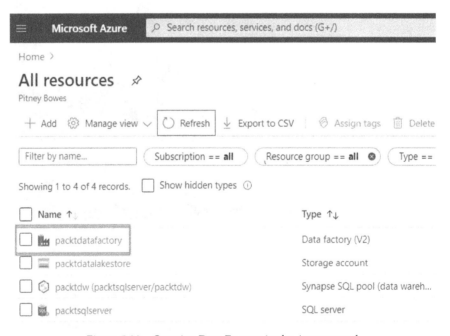

Figure 5.28 – Opening Data Factory in the Azure portal

6. On the **packtdatafactory** overview page, select **Author & Monitor**:

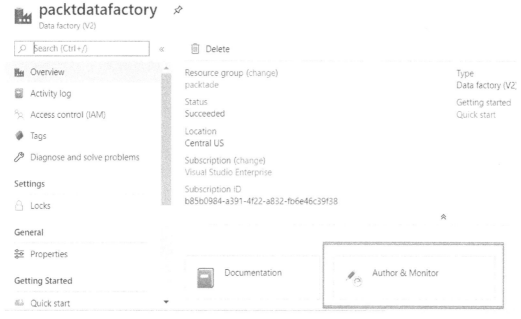

Figure 5.29 – Selecting Author & Monitor

7. A new browser tab will open. In the new browser tab, select **Create pipeline**:

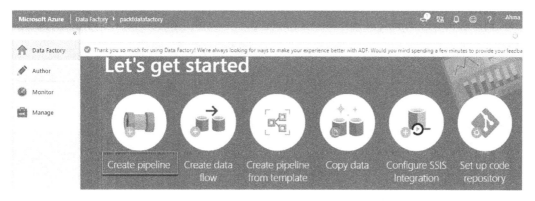

Figure 5.30 – Selecting Create pipeline

8. On the **packtdatafactory** page, change the pipeline name and description, as shown in the following screenshot:

Figure 5.31 – Changing the pipeline name and description

9. We'll now create the linked service for the source and destination. A linked service is a connection to the source and destination. To create a linked service to the Azure Data Lake Storage Gen2 account, select the **Manage** tab from the left menu:

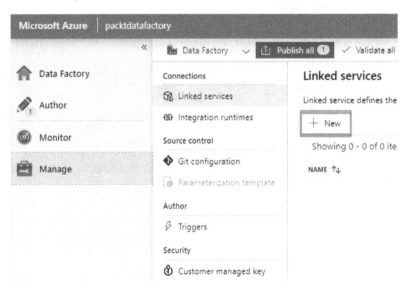

Figure 5.32 – Opening the Manage tab

10. Click on the **+ New** link to create a new linked service. On the **New linked service** page, type `data lake` in the **Search** box and select **Azure Data Lake Storage Gen2**:

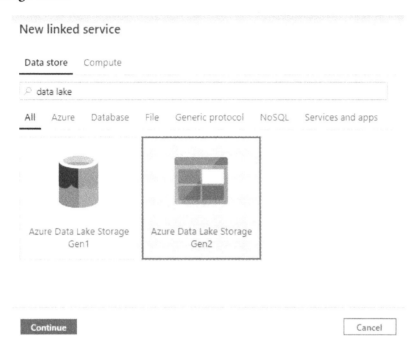

Figure 5.33 – Selecting the Azure Data Lake Storage Gen2 linked service

Click **Continue** to move to the next page.

11. On the next page, provide the linked service name and a description (optional) and set the authentication method as **Account key**. Select the subscription that contains the Data Lake store, and then select the Data Lake Storage account under **Storage account name**:

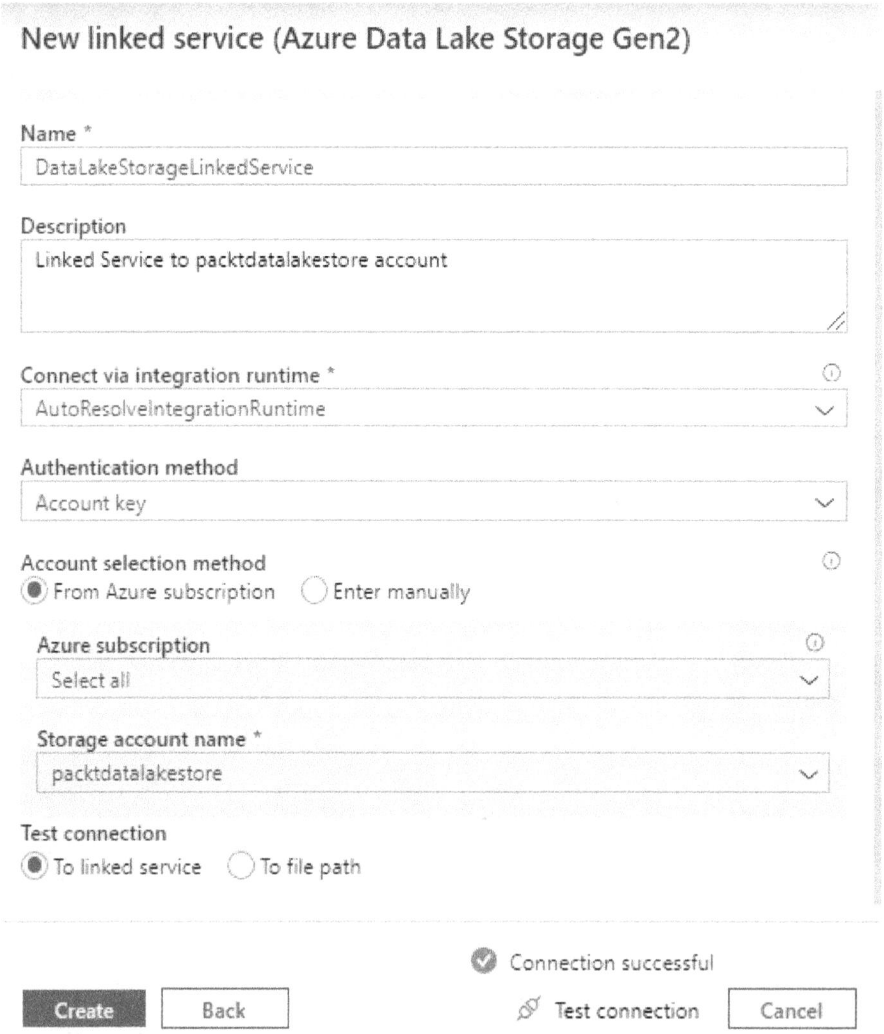

Figure 5.34 – Creating an Azure Data Lake Storage Gen2 linked service

Select **Test connection** to verify that the linked service is able to connect to the given Data Lake Storage account. If the connection is successful, click **Create** to create the linked service.

12. We'll now create the linked service to connect to the Azure Synapse SQL pool. Click **+ New link** on the **Linked services** page to create a new linked service. On the **New linked service** page, type Synapse in the **Search** box and select **Azure Synapse Analytics**:

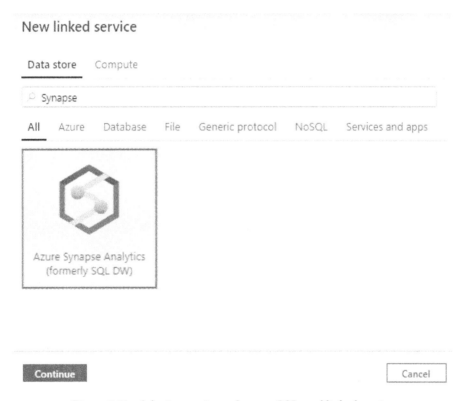

Figure 5.35 – Selecting an Azure Synapse SQL pool linked service

Click **Continue** to move to the next page.

13. On the **New linked service** page, provide the linked service name and description (optional). Select **From Azure subscription** for **Account selection method,** and then select the Azure SQL server from the **Server name** drop-down list. Select the SQL pool from the **Database name** drop-down list. Provide the SQL admin username and password:

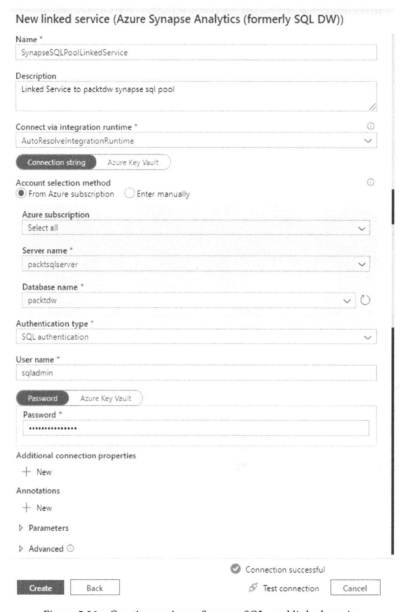

Figure 5.36 – Creating an Azure Synapse SQL pool linked service

Click **Test connection** to verify the connectivity to the Synapse SQL pool. If the connection is successful, click **Create** to create the linked service. When the linked service is created, we'll be taken back to the **Linked services** page. We can see the linked services listed on the **Linked services** page:

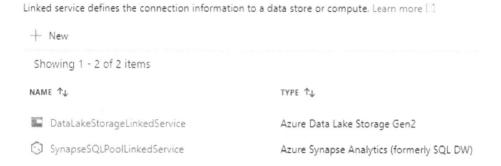

Figure 5.37 – Viewing linked services

14. The next step is to create a pipeline and configure the copy activity. On the **packtdatafactory** page, select **Author** from the left menu. Select the + icon beside the **Search** box and then **Pipeline** to create a new pipeline:

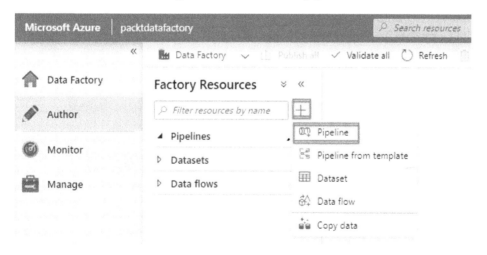

Figure 5.38 – Creating a new Azure Data Factory pipeline

15. Specify the pipeline name as `Pipeline-ADL-SQLPool` and add a pipeline description in the **Properties** pane, as shown in the following screenshot:

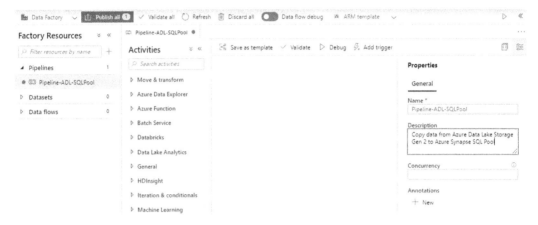

Figure 5.39 – Modifying the pipeline name and description

> **Note**
>
> After specifying the pipeline name and description, you can close the **Properties** pane by clicking on the right icon at the top of the **Properties** pane.

16. Under the **Activities** menu, expand **Move & transform** and drag and drop the **Copy data** activity onto the pipeline canvas. Under the **General** tab, change the name of the **Copy data** activity to `Copy Orders from Data Lake to SQL Pool`:

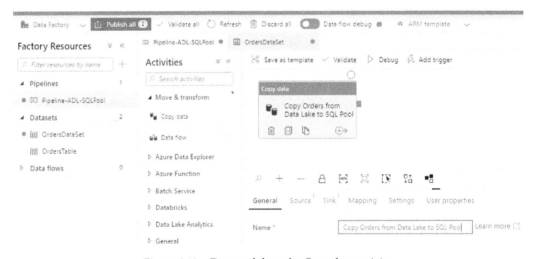

Figure 5.40 – Drag and drop the Copy data activity

17. The next step is to configure the copy data activity source. To do this, select the **Source** tab from the pipeline canvas:

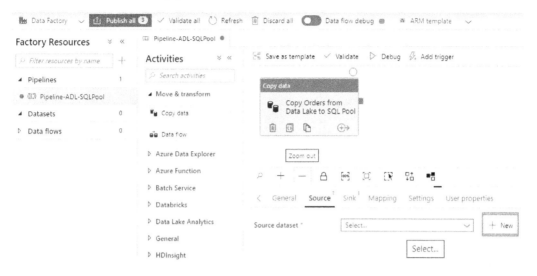

Figure 5.41 – Creating a new source dataset

18. Click on the **+ New** button to create a new dataset. On the **New dataset** page, type `data lake` in the **Search** box, and then select **Azure Data Lake Storage Gen2**:

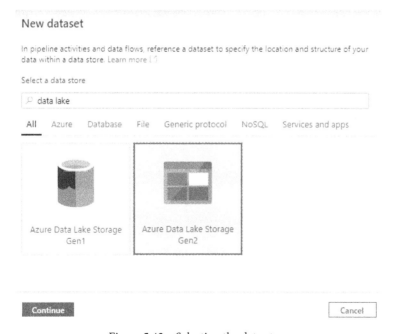

Figure 5.42 – Selecting the data store

19. Click **Continue** to go to the next page. On the **Select format** page, select **DelimitedText**:

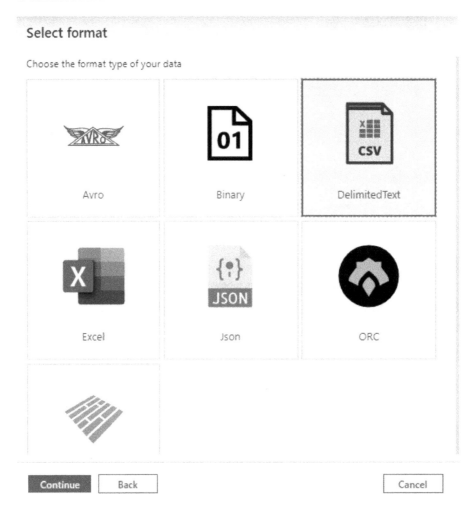

Figure 5.43 – Selecting a dataset file format

Click **Continue** to go to the next page.

20. On the **Set properties** page, provide the dataset name as `OrdersFile` and select **DataLakeStorageLinkedService** from the **Linked service** drop-down menu. Provide the **File path** value as `ecommerce/orders/orders1.txt`:

Set properties

Name

| OrdersFile |

Linked service *

| DataLakeStorageLinkedService | ⌄ | 🖉 |

> Filename doesn't support wildcard in dataset

File path

| ecommerce | / | orders | / | orders1.txt | 🗁 | ⌄ |

First row as header ☐

Import schema

◉ From connection/store   ○ From sample file   ○ None

▷ Advanced

| **OK** | Back | | Cancel |

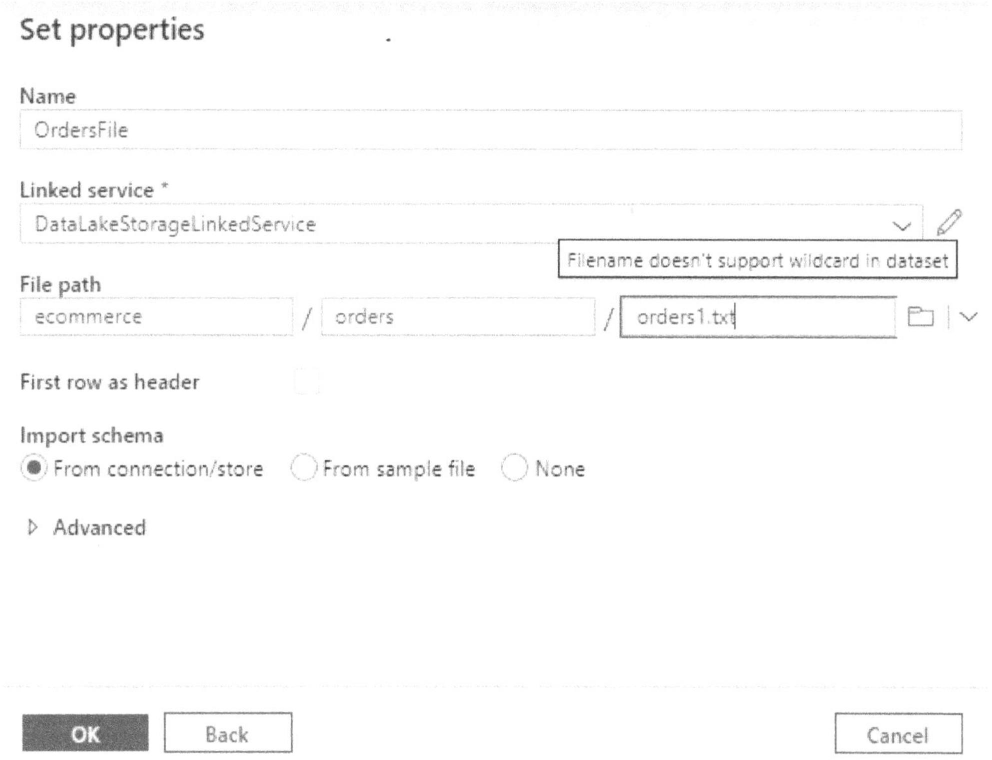

Figure 5.44 – Specifying the dataset properties

> **Note**
> The file path specifies the filesystem, directory, and the file in the given Data Lake Storage account.

Click **OK** to create the dataset. Once the dataset is created, we are taken back to the pipeline canvas. Let's now view and modify the dataset settings.

21. Under the **Factory Resources** menu, expand **Datasets** and select the **OrdersFile** dataset:

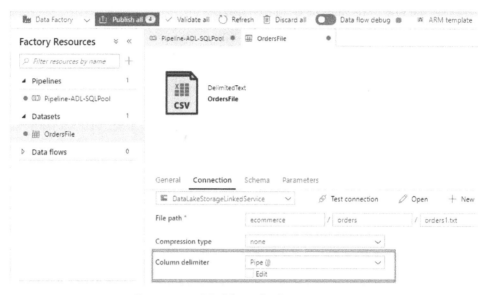

Figure 5.45 – Modifying the dataset settings

Select the **Connection** tab and change the **Column delimiter** value from **Comma** to **Pipe**.

22. Scroll to the right and find and click on the **Preview data** button:

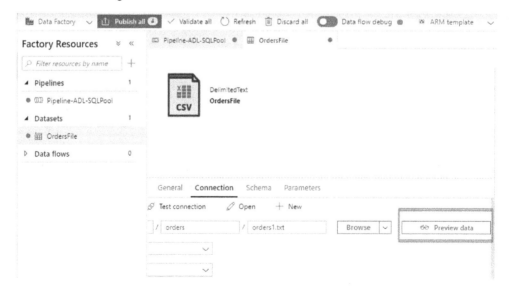

Figure 5.46 – Previewing data in the dataset

A new preview window will open with the data preview:

Preview data

Linked service: DataLakeStorageLinkedService

Object: orders1.txt

Prop_0	Prop_1	Prop_2	Prop_3	Prop_4	Prop_5	Prop_6	Prop_7	Prop_8
540468	17012A	ORIGAMI VANILLA INCENSE/CANDLE SET	1	2011-01-07 13:55:00	2.51		United Kingdom	1
540468	18097C	WHITE TALL PORCELAIN T-LIGHT HOLDER	3	2011-01-07 13:55:00	5.06		United Kingdom	2
540468	20615	BLUE POLKADOT PASSPORT COVER	2	2011-01-07 13:55:00	1.66		United Kingdom	3
540468	20652	BLUE POLKADOT LUGGAGE TAG	2	2011-01-07 13:55:00	0.85		United Kingdom	4
540468	20653	CHERRY BLOSSOM LUGGAGE TAG	1	2011-01-07 13:55:00	0.85		United Kingdom	5
540468	20699	MOUSEY LONG LEGS SOFT TOY	1	2011-01-07 13:55:00	2.51		United Kingdom	6
540468	20717	STRAWBERRY SHOPPER BAG	14	2011-01-07	0.85		United	7

Figure 5.47 – Dataset preview

Close the window. The source for our copy data activity is now ready.

23. To configure the sink for the copy data activity, under the **Factory Resources** menu, expand **Pipelines** and select **Pipeline-ADL-SQLPool**. In the pipeline canvas, select the **Sink** tab of the copy data activity:

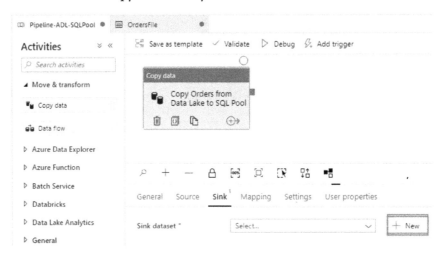

Figure 5.48 – Creating a sink dataset

24. Click the **+ New** button to create a new sink dataset. On the **New dataset** page, type synapse in the **Search** box, and then select **Azure Synapse Analytics**:

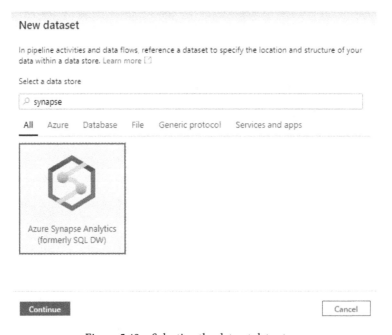

Figure 5.49 – Selecting the dataset data store

Click **Continue** to go to the next step.

25. On the **Set properties** page, change the dataset name to `OrdersTable`, select **SynapseSQLPoolLinkedService** from the **Linked service** dropdown, and select the **dbo.Orders** table from the **Table name** dropdown:

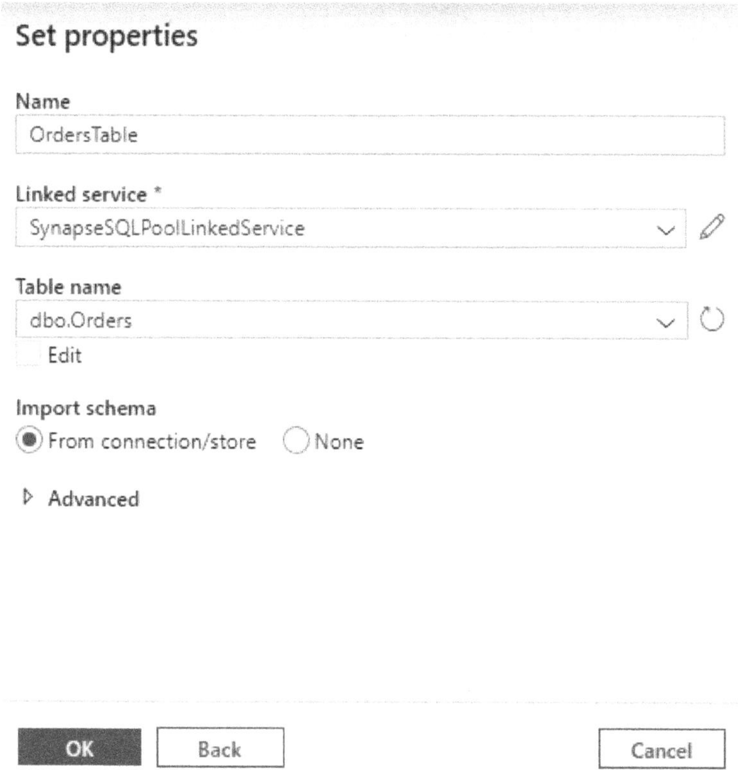

Set properties

Name

OrdersTable

Linked service *

SynapseSQLPoolLinkedService

Table name

dbo.Orders

Edit

Import schema

⦿ From connection/store   ◯ None

▷ Advanced

OK   Back   Cancel

Figure 5.50 – Configuring the dataset properties

Click the **OK** button to create the dataset. The sink dataset is now created:

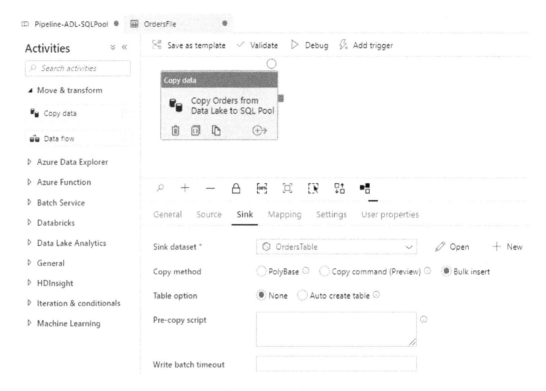

Figure 5.51 – Viewing the sink dataset settings

There are three different copy methods to copy data to a Synapse SQL pool: **PolyBase**, **Copy command**, and **Bulk insert**. The default is **PolyBase**, but we'll use **Bulk insert** for our example.

26. We'll modify a few of the sink dataset properties and configure the mapping between the source and sink datasets. In the pipeline canvas, click on the copy data activity, and then select the **Sink** tab. Then, copy and paste the following T-SQL script into the **Pre-copy script** section:

```
TRUNCATE TABLE dbo.Orders
```

The screenshot for reference is as follows:

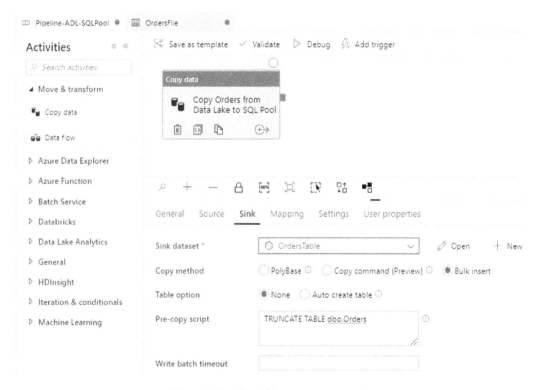

Figure 5.52 – Specifying a pre-copy script

> **Note**
>
> We can use the **Auto create table** option to create the table automatically; however, in that case, we need to specify the mappings between the source and the sink dataset manually.

27. To map the source and sink datasets, select the **Mapping** tab. Click on the **Import schemas** button to import the source and sink dataset schema. Modify the datatype for the **Quantity** and **orderid** columns to **Int64** and **UnitPrice** to **Decimal**:

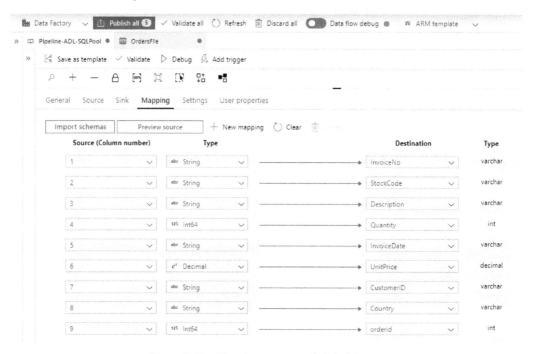

Figure 5.53 – Mapping source and sink datasets

The copy data activity is now configured and our pipeline is ready to be executed. Click **Publish all** in the top menu to save the pipeline:

Figure 5.54 – Publishing the pipeline

28. To test the pipeline, click **Debug** from the top menu:

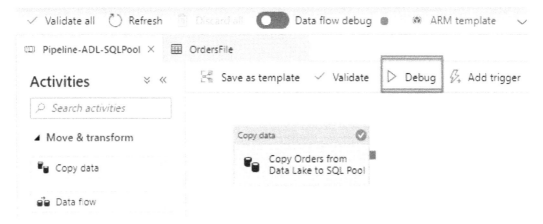

Figure 5.55 – Testing the pipeline

When the pipeline completes, the result is displayed on the bottom pane:

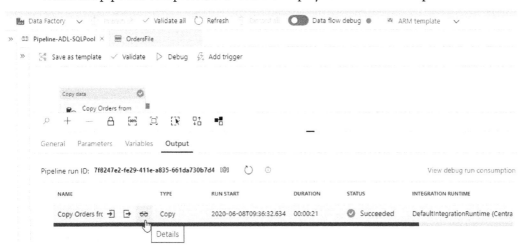

Figure 5.56 – Viewing the execution output

To view the execution details, hover the mouse over the copy data activity name and click on the spectacles icon. You should get similar details as shown in the following screenshot:

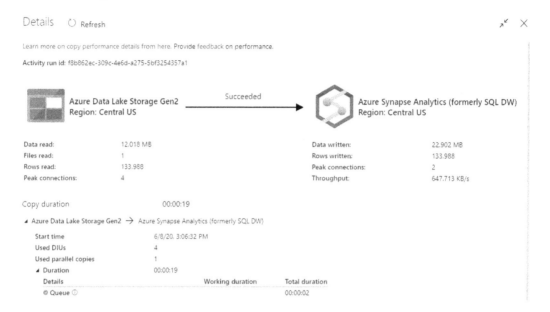

Details　◯ Refresh

Learn more on copy performance details from here. Provide feedback on performance.

Activity run id: f8b862ec-309c-4e6d-a275-5bf3254357a1

Azure Data Lake Storage Gen2　　　Succeeded　　　Azure Synapse Analytics (formerly SQL DW)
Region: Central US　　　　　　　　　　　　　　　　　Region: Central US

Data read:	12.018 MB	Data written:	22.902 MB
Files read:	1	Rows written:	133.988
Rows read:	133.988	Peak connections:	2
Peak connections:	4	Throughput:	647.713 KB/s

Copy duration　　　　　　　　00:00:19

▴ Azure Data Lake Storage Gen2 → Azure Synapse Analytics (formerly SQL DW)

Start time	6/8/20, 3:06:32 PM
Used DIUs	4
Used parallel copies	1
▴ Duration	00:00:19

Details	Working duration	Total duration
◎ Queue ⓘ		00:00:02

Figure 5.57 – Viewing execution details

Observe that the pipeline copied 133,988 records from orders1.txt to the Orders table in the Azure Synapse SQL pool.

Copying data from Azure Data Lake Gen2 to Azure Cosmos DB using the copy activity

In this recipe, we'll copy data from Azure Data Lake Gen2 to an Azure Cosmos DB SQL API. Azure Cosmos DB is a managed NoSQL database service and offers multiple NoSQL databases, such as MongoDB, DocumentDB, GraphDB (Gremlin), Azure Table storage, and Cassandra, to store data.

Getting ready

Before you start, do the following:

1. Log in to Azure from PowerShell. To do this, execute the following command and follow the instructions to log in to Azure:

    ```
    Connect-AzAccount
    ```

2. Open https://portal.azure.com and log in using your Azure credentials.

3. Follow *step 1* of the *Copying data from Azure Data Lake Gen2 to an Azure Synapse SQL pool using the copy activity* recipe to create and upload files to Azure Data Lake Storage Gen2.

To copy data from Azure Data Lake Storage Gen2 to a Cosmos DB SQL API, we'll do the following:

1. Create and upload data to the Azure Data Lake Storage Gen2 account. To do this, follow *step 1* of the *Copying data from Azure Data Lake Gen2 to an Azure Synapse SQL pool using the copy activity* recipe.

2. Create a new Cosmos DB account and a SQL database.

3. Create an Azure data factory if one doesn't exist.

4. Create a linked service for Azure Data Lake Storage Gen2.

5. Create a linked service for Azure Cosmos DB.

6. Create a pipeline and configure the copy activity.

How to do it...

Let's get started with creating a new Cosmos DB account and SQL database:

1. Execute the following PowerShell command to create a new Cosmos DB account (packtcosmosdb), Cosmos DB SQL API database (orders), and container (weborders):

    ```
    .\ADE\azure-data-engineering-cookbook\Chapter04\4_
    CreatingCosmosDBSQLDatabase.ps1 -resourceGroupName
    packtade -location centralus -accountName packtcosmosdb
    -apiKind Sql
    ```

You should get an output as shown:

Figure 5.58 – Creating an Azure Cosmos DB account and database

Note the connection string as it's required when creating the linked service.

2. Let's add a sample item to the container. To do this, in the Azure portal, open **All resources** and select the packtcosmosdb account. On the **packtcosmosdb** page, select **Data Explorer**. Under the **SQL API** header, expand the orders database and then the weborders container, and then select **Items**. Click on **New Item** from the top menu, and then copy and paste the following to create a new item:

```
{
    "InvoiceNo": "540468",
    "StockCode": "17012A",
    "Description": "ORIGAMI VANILLA INCENSE/CANDLE SET",
    "Quantity": "1",
    "InvoiceDate": "2011-01-07 13:55:00",
    "UnitPrice": "2.51",
    "CustomerID": "12312",
    "Country": "United Kingdom",
    "orderid": "1",
    "createTime": "2018-05-21T22:59:59.999Z"
}
```

The screen will look as follows:

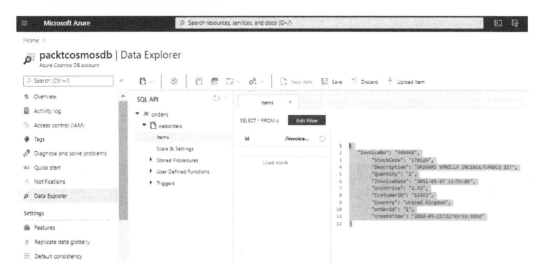

Figure 5.59 – Creating a new item in the container

Click **Save** to save the record.

3. Execute the following PowerShell script to create an Azure Data Factory account. If you created an account in an earlier recipe, you can skip this step:

```
.\ADE\azure-data-engineering-cookbook\Chapter04\3_
CreatingAzureDataFactory.ps1 -resourcegroupname packtade
-location centralus -datafactoryname packtdatafactory
```

4. To create a linked service for Cosmos DB, open the ~/
    Chapter04/CosmosDBLinkedServiceDefinitionFile.txt file
    in Notepad. Replace the AccountEndpoint and AccountKey values in
    the connectionString section with what was obtained in *step 1*. The
    connectionString section should be as shown:

```
"connectionString": "AccountEndpoint=https://
packtcosmosdb.
documents.azure.com:443/;Database=Orders;AccountKey=
wQJTCv6KMcY99kf0vhSknPftUMQmjWOwbkUrDd1riMbEkAyJQRcRDhm
OBzd6s8vqrERpZdyiqEpYIfmwkJ23og==;"
```

5. Save the file, and then execute the following PowerShell command to create the Cosmos DB linked service:

```
Set-AzDataFactoryV2LinkedService -Name
CosmosDbLinkedService -DefinitionFile C:\ade\
azure-data-engineering-cookbook\Chapter04\
CosmosDBLinkedServiceDefinitionFile.txt
-ResourceGroupName packtade -DataFactoryName
packtdatafactory
```

You should get an output as shown:

```
PS C:\> Set-AzDataFactoryV2LinkedService -Name CosmosDbLinkedService -DefinitionFile C:\ad
e\azure-data-engineering-cookbook\Chapter6\CosmosDBLinkedServiceDefinitionFile.txt -Resour
ceGroupName packtade -DataFactoryName packtdatafactory

LinkedServiceName : CosmosDbLinkedService
ResourceGroupName : packtade
DataFactoryName   : packtdatafactory
Properties        : Microsoft.Azure.Management.DataFactory.Models.CosmosDbLinkedService
```

Figure 5.60 – Creating an Azure Cosmos DB linked service

6. To create a linked service for Azure Data Lake Storage Gen2, follow *step 4* of the *Copying data from Azure Data Lake Gen2 to an Azure Synapse SQL pool using the copy activity* recipe.

7. In the Azure portal, under **All resources**, find and open packtdatafactory. On the **Overview** page, click on **Author & Monitor**. On the **Azure Data Factory** page, select **Create pipeline**.

> **Note**
>
> Screenshots are not included for the sake of brevity. Check *step 4* of the *Copying data from Azure Data Lake Gen2 to an Azure Synapse SQL pool using the copy activity* recipe for details.

8. Name the pipeline Pipeline-ADL-CosmosSQLAPI and provide an optional description of Copy data from Azure Data Lake Storage Gen 2 to Azure Cosmos DB SQL API:

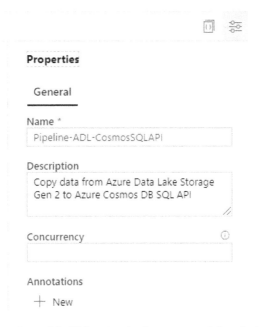

Figure 5.61 – Modifying the pipeline name and description

9. Under the **Activities** menu, expand **Move & transform** and drag and drop the
    Copy data activity onto the canvas. Rename the activity to `Copy Orders from
    Data Lake to Cosmos DB`. Configure the source as specified in *step 8*.

10. To configure the sink, select the copy data activity and select the **Sink** tab:

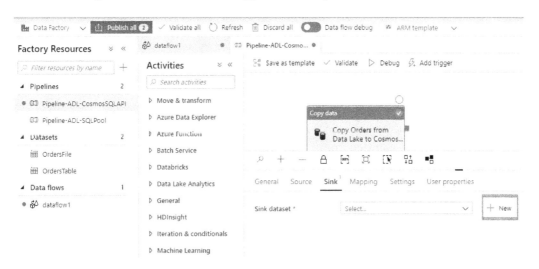

Figure 5.62 – Configuring the copy data activity sink

Click **+ New** to add a new dataset.

11. On the **New dataset** page, type cosmos in the **Search** box and select **Azure Cosmos DB (SQL API)**:

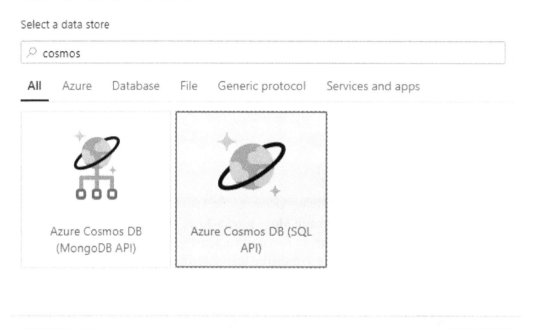

Figure 5.63 – Selecting the sink data store

Click **Continue** to go to the next page.

12. On the **Set properties** page, set the dataset name as Ordersdb, select **CosmosDbLinkedService** from the **Linked service** dropdown, and select **weborders** from the **Collection** dropdown:

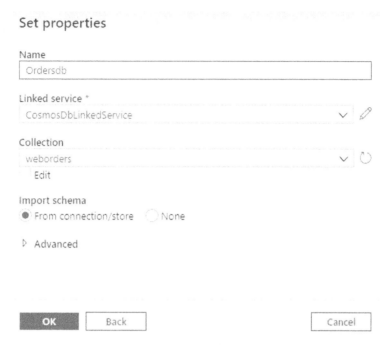

Figure 5.64 – Setting the dataset properties

Click **OK** to create the new dataset.

13. We'll be taken back to the pipeline canvas. Observe that the new dataset, Ordersdb, is selected:

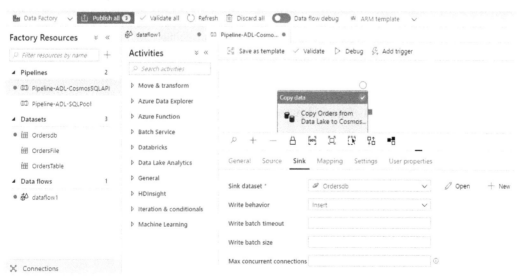

Figure 5.65 – Viewing the dataset properties

14. The next step is to map the source and sink columns. To map the columns, select the **Mapping** tab and click **Import schemas**:

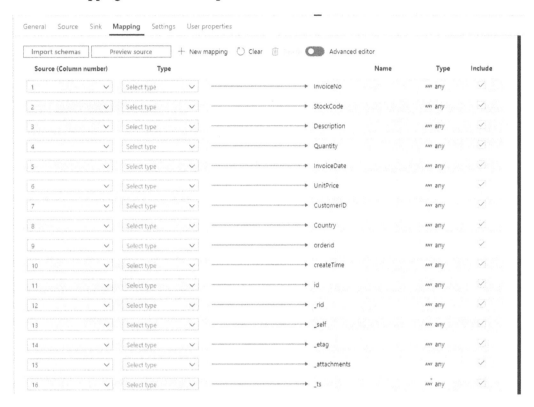

Figure 5.66 – Importing the dataset mappings

Observe that there are more columns for the source, such as id, _rid, _self, _etag, _attachments, and _ts. These are auto-generated columns and we don't need to insert these. So, we'll exclude these columns by unchecking the checkbox against each of them. Then, modify the datatype for the columns as listed:

a) InvoiceNo – **String**

b) StockCode – **String**

c) Description – **String**

d) Quantity – **Int64**

e) InvoiceDate – **String**

f) UnitPrice – **Decimal**

g) CustomerID – **String**

h) Country – **String**

i) orderid – **Int64**

The following is a screenshot for reference:

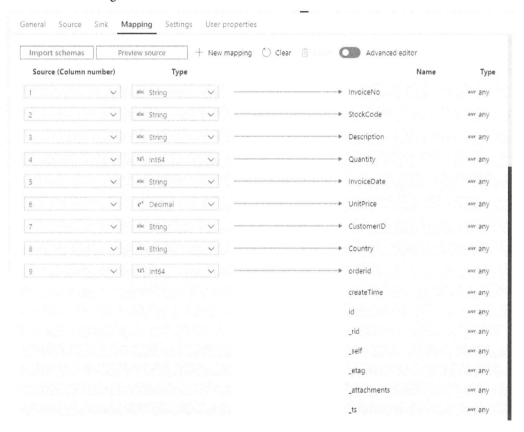

Figure 5.67 – Configuring the dataset mappings

15. This completes the copy data activity configuration. Click on the **Publish all** button to save the pipeline:

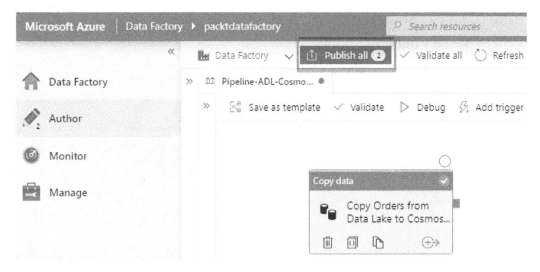

Figure 5.68 – Saving the pipeline

16. Click on the **Debug** button to run the pipeline.

17. To view the progress, select the copy data activity and select the **Output** tab. Hover the mouse over the text under the column name and click on the spectacles icon:

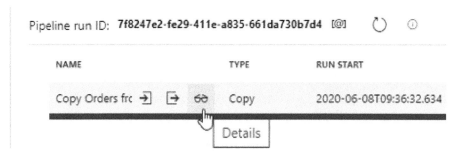

Figure 5.69 – Viewing the pipeline progress

Note
It may take 10–12 minutes for the pipeline to complete.

When the pipeline completes, the status is updated as shown:

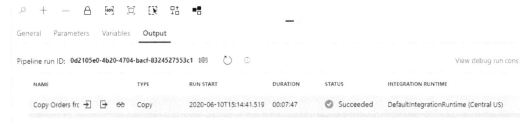

Figure 5.70 – Viewing the pipeline execution status

18. To verify the data in Cosmos DB, open `packtcosmosdb` and then select **Data Explorer**. Under the **SQL API** tab, expand the `orders` database and the `weborders` container. Select **New SQL Query** from the top menu. Copy and paste the following query in the new query window, and then click **Execute Query**:

```
SELECT VALUE COUNT(1) FROM c
```

You should get an output as shown:

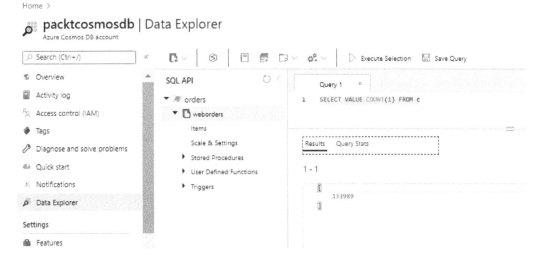

Figure 5.71 – Verifying data in Cosmos DB

Execute the following command to delete the Cosmos DB account:

```
Remove-AzCosmosDBAccount -ResourceGroupName packtade
-Name packtcosmosdb
```

> **Note**
> If you have used a different Cosmos DB account name and resource group, modify the parameter values in the preceding command accordingly.

6
Data Flows in Azure Data Factory

In this chapter, we'll look at two data flow activities: the mapping data flow and the wrangling data flow. Data flow activities provide a code-free way to implement transformations on the fly as and when data is being processed.

Incremental data loading is a very common scenario wherein data from a source is incrementally loaded to a destination. There are multiple ways to implement incremental data flows. This chapter provides an implementation of incremental data loading that you can use in your environment as and when required.

The wrangling data flow provides a code-free UI that can be used to clean and transform data using Power Query. This makes it easy for non-developers to implement data transformation and cleaning and create data pipelines quickly.

In this chapter, we'll cover the following recipes:

- Implementing incremental data loading with a mapping data flow
- Implementing a wrangling data flow

> **Note**
> The wrangling data flow is in preview at the time of writing this book.

Technical requirements

For this chapter, the following are required:

- A Microsoft Azure subscription
- PowerShell 7
- Microsoft Azure PowerShell

Implementing incremental data loading with a mapping data flow

A mapping data flow provides a code-free data flow transformation environment. We use the UI to implement the ETL and process the pipeline. Spark clusters are then provisioned, and then the data flow is transformed to Spark code and executed.

In this recipe, we'll look at one of the approaches to implement incremental data loading using a mapping data flow.

Getting ready

To get started, do the following:

1. Log in to https://portal.azure.com using your Azure credentials.

2. Open a new PowerShell prompt. Execute the following command to log in to your Azure account from PowerShell:

```
Connect-AzAccount
```

3. You will need an existing Data Factory account. If you don't have one, create one by executing the following PowerShell script: ~/azure-data-engineering-cookbook\Chapter04\3_CreatingAzureDataFactory.ps1.

4. Create an Azure storage account and upload files to the ~/Chapter06/Data folder in orders/datain containers. You can use the ~/Chapter05/1_UploadDatatoAzureStorage.ps1 PowerShell script to create the storage account and upload the files.

How to do it...

In this recipe, we'll use a mapping data flow to process orders. The orders.txt file is a pipe-delimited list of orders available on Azure Storage. We'll use a mapping data flow to read the orders.txt file, and then we'll transform and aggregate the data to calculate the sales by customer and country. We'll then insert the data into a table in Azure SQL Database and the /orders/dataout folder in Azure Storage.

Let's start by creating the Azure SQL database and the required database objects:

1. Execute the following PowerShell command to create the Azure SQL database:

    ```
    .\azure-data-engineering-cookbook\Chapter06\1_
    ProvisioningAzureSQLDB.ps1 -resourcegroup packtade
    -servername azadesqlserver -databasename azadesqldb
    -password Sql@Server@1234 -location centralus
    ```

    The preceding command creates the packtade resource group if it doesn't exist, and an Azure SQL database, azadesqldb.

2. Connect to the Azure SQL database either using **SQL Server Management Studio** (**SSMS**) or the Azure portal and execute the following queries to create the Customer and Sales tables:

    ```
    CREATE TABLE customer
      (
         id INT,
         [name] VARCHAR(100)
      )
    GO
    INSERT INTO customer
      VALUES(1,'Hyphen'),(2,'Page'),(3,'Data
    Inc'),(4,'Delta'),(5,'Genx'),(6,'Rand Inc'),(7,'Hallo
    Inc')
    GO
    CREATE TABLE Sales
      (
         CustomerName VARCHAR(100),
         Country VARCHAR(100),
         Amount DECIMAL(10,2),
         CreateDate DATETIME DEFAULT (GETDATE())
      )
    ```

```
GO
CREATE TABLE SalesStaging
  (
     CustomerName VARCHAR(100),
     Country VARCHAR(100),
     Amount DECIMAL(10,2)
  )
GO
CREATE PROCEDURE dbo.MergeSales
AS
BEGIN
     SET NOCOUNT ON;
     MERGE Sales AS target
     USING SalesStaging AS source
     ON (target.Country = source.Country and target.
CustomerName=source.CustomerName)
     WHEN MATCHED THEN
        UPDATE SET Amount = source.Amount
     WHEN NOT MATCHED THEN
        INSERT (CustomerName,Country,Amount)
        VALUES (source.CustomerName, source.
Country,source.Amount);

END;
```

3. We'll now create the linked service for the Azure SQL database. Open ~/
    Chapter06/AzureSQLDBLinkedServiceDefinition.txt. Modify
    the connection string as required. Save and close the file. Execute the following
    PowerShell command to create the linked service:

```
Set-AzDataFactoryV2LinkedService -Name
AzureSqlDatabaseLinkedService -DefinitionFile .\
azure-data-engineering-cookbook\Chapter06\
AzureSQLDBLinkedServiceDefinition.txt -ResourceGroupName
packtade -DataFactoryName packtdatafactory
```

4. We'll now create a linked service to the Azure storage account that has the `orders.txt` file. Open `~/Chapter06/PacktStorageLinkedServiceDefinition.txt`. Provide the values for the storage account name and account key. Save and close the file. Execute the following command to create the linked service:

```
Set-AzDataFactoryV2LinkedService -Name
PacktStorageLinkedService -DefinitionFile .\
azure-data-engineering-cookbook\Chapter06\
PacktStorageLinkedServiceDefinition.txt -DataFactoryName
packtdatafactory -ResourceGroupName packtade
```

5. We'll now create a dataset for the `orders.txt` file in Azure Blob storage. To do that, open `~/Chapter06/OrdersDatasetDefinition.txt`. Modify the container name and folder path if it's different in your case. Save and close the file. Execute the following command to create the dataset:

```
New-AzDataFactoryV2Dataset -Name orders -DefinitionFile
.\azure-data-engineering-cookbook\Chapter06\
OrdersDatasetDefinition.txt -ResourceGroupName packtade
-DataFactoryName packtdatafactory
```

The preceding command uses the `New-AzDataFactoryV2Dataset` cmdlet to create the dataset for `PacktStorageLinkedService`. You may have to change the resource group and data factory name according to your environment.

6. Similarly, open `~/Chapter06/CustomerDatasetDefinition.txt` and `~/Chapter06/SalesDatasetDefinition.txt` and modify the linked service name if it's different. Save and close the files. Execute the following commands to create the `CustomerTable` and `SalesStagingTable` datasets for the `Customer` and `Sales` tables in the Azure SQL database:

```
New-AzDataFactoryV2Dataset -Name CustomerTable
-DefinitionFile .\azure-data-engineering-cookbook\
Chapter06\CustomerDatasetDefinition.txt -ResourceGroupName
packtade -DataFactoryName packtdatafactory
```

```
New-AzDataFactoryV2Dataset -Name SalesStagingTable
-DefinitionFile .\azure-data-engineering-
cookbook\Chapter06\SalesStagingDatasetDefinition.
txt -ResourceGroupName packtade -DataFactoryName
packtdatafactory
```

```
New-AzDataFactoryV2Dataset -Name RunningTotalTable
-DefinitionFile .\azure-data-engineering-
cookbook\Chapter06\RunningTotalDatasetDefinition.
txt -ResourceGroupName packtade -DataFactoryName
packtdatafactory
```

7. We'll now start developing the pipeline. In the Azure portal, open the **Data Factory Author** page and create a new pipeline named Pipeline-Mapping-DataFlow. Drag and drop the **Data flow** activity under **Move & transform** on the **Activities** tab on the pipeline canvas. On the **Adding data flow** page, select **Create new data flow** and then select **Mapping Data Flow**:

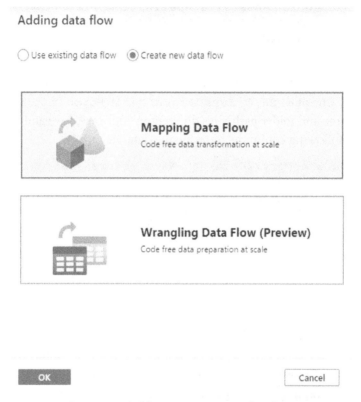

Figure 6.1 – Adding a new mapping data flow

8. Click **OK** to continue. A new data flow tab is available in the data factory named dataflow1.

9. We'll start by adding the source to the data flow. To do that, click on **Add source text**. Under **Source settings**, rename the output stream ordersraw.

    From the **Dataset** dropdown, select the orders dataset. Leave the rest of the options as they are:

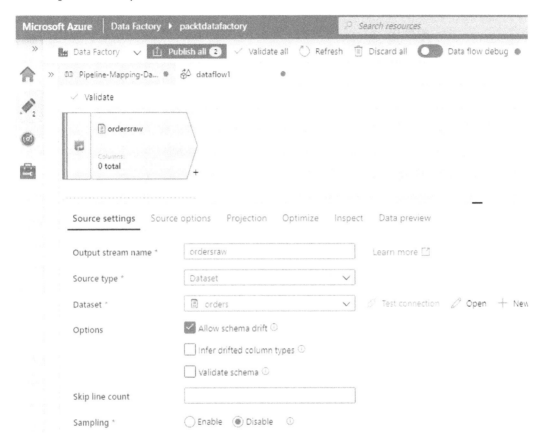

Figure 6.2 – Configuring the source – Source settings

10. Switch to the **Source options** tab. Select **Move** for the **After completion** action and add `processedfiles` in the **To** text box. The **From** text box should already have the input path orders, `/datain`. Set **Column to store file name** to `datafilename`. Now, `datafilename` will add a new column in the data stream that contains the file being processed:

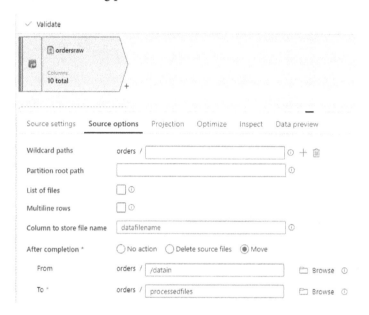

Figure 6.3 – Configuring Source options

11. Switch to the **Projection** tab. In this tab, we can import the file schema. However, to import the schema and preview the data, we click on the **Data flow debug** switch and then click **OK** in the **Turn on data flow debug** dialog box to enable debug mode. This can take up to 10-15 minutes. When the data flow debug mode is turned on, click on **Import projection**. The file schema will be imported. The `orders.txt` file doesn't have headers, so column names will be assigned as `_col1_`, `_col2_`, and so on. Modify the column names as shown in the following screenshot:

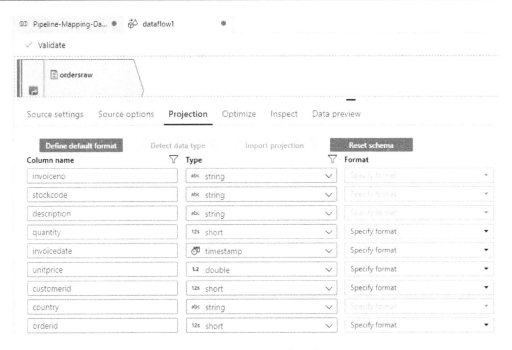

Figure 6.4 – Importing the schema

12. Switch to the **Data preview** tab and click **Refresh** to preview the data. You should get a similar output to what is shown in the following figure:

Figure 6.5 – Previewing data

Observe that the `datafilename` column contains the source file path. This helps in reconciling whenever required.

13. The `orders` data in *step 11* has `CustomerId` and not `CustomerName`. Therefore, we'll join the customer table from Azure SQL Database. To do this, click on the **Add Source** text below the `ordersraw` source. Under the **Source settings** tab, provide the output stream name `Customer`. From the **Dataset** dropdown, select the `CustomerTable` dataset:

Figure 6.6 – Configuring Source settings for Customer Source

> **Note**
> The **Source options** tab allows you to select the data from the entire table or from a query. We need the data from the table; therefore, we can leave the values under the **Source options** tab as their defaults.

14. Switch to the **Data preview** tab and hit **Refresh** to view the data:

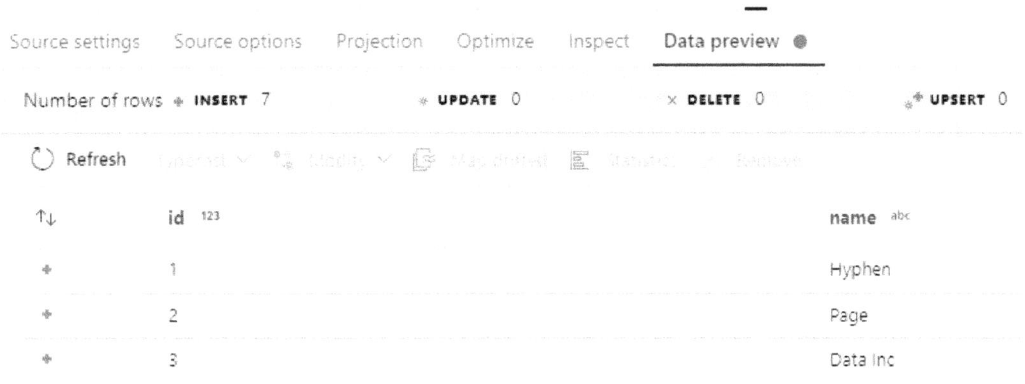

	id 123		name abc

Source settings   Source options   Projection   Optimize   Inspect   **Data preview** ●

Number of rows ✛ **INSERT** 7   ✳ **UPDATE** 0   ✕ **DELETE** 0   ✚ **UPSERT** 0

◯ Refresh

↑↓	id 123		name abc
✛	1		Hyphen
✛	2		Page
✛	3		Data Inc

Figure 6.7 – Previewing the customer data

15. Select the + icon at the bottom of the `ordersraw` box and select **Join**:

Figure 6.8 – Joining ordersraw and the customer source

16. The **Join1** transformation box will be added to the canvas. Under the **Join settings** tab, rename the output stream `JoinOrdersCustomer`. Set **Left stream** to `ordersraw` and **Right stream** to `Customer`. Set **Join type** to **Inner**. Under **Join conditions | Left: ordersraw's column**, select **customerid**, and for **Right: Customer's column**, select **id**:

Figure 6.9 – Configuring the join transformation

Switch to the **Data preview** tab and hit **Refresh**:

Figure 6.10 – Viewing the join transformation output

Observe that the data stream now has two new columns from the `Customer` table, `id` and `name`.

17. We'll now add a derived column transformation to add a new column, Amount. The Amount column is the product of quantity and unit price. To do that, click on the + icon at the bottom of the **JoinOrdersCustomer** transform and then select **Derived Column** from the context menu. A new **DerivedColumn1** transformation is added to the data flow. Under the **Derived column's settings** tab, rename the output stream name to DerivedAmount. Under the **Columns** section, type Amount in the text box where it says **Add or select a column**. Click on the **Enter expression** text and then copy and paste quantity*unitprice in the **Visual Expression Builder** window. Click **Save and close**:

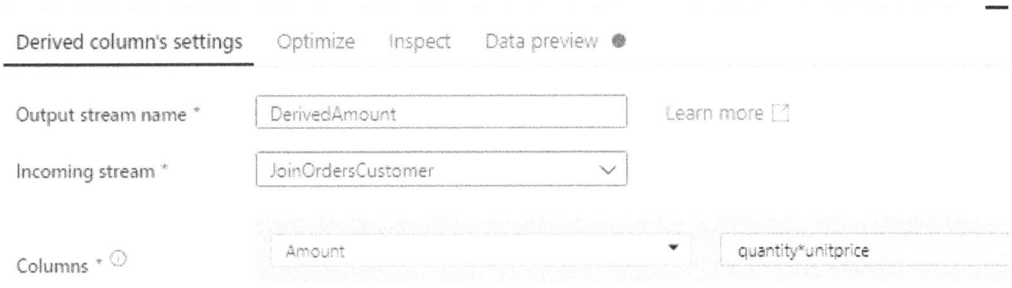

| Derived column's settings | Optimize | Inspect | Data preview ● |

Output stream name *	DerivedAmount	Learn more ☐
Incoming stream *	JoinOrdersCustomer ⌄	
Columns * ⓘ	Amount ▾	quantity*unitprice

Figure 6.11 – Configuring the derived column transformation

Under the **Data preview** tab, hit **Refresh** to view the data. A new Amount column is added to the data stream.

18. We'll now add a filter transformation to filter out the rows where the country is unspecified. To do that, click the + icon at the bottom of the **DerivedAmount** transformation. Select the filter transformation from the dropdown. Under the **Filter settings** tab, name the output stream name FilterCountry. Click on the **Enter filter text** area. Enter country!='Unspecified' in the visual expression builder. Click **Save and finish** to close the visual expression builder:

| Filter settings | Optimize | Inspect | Data preview ● |

Output stream name *	FilterCountry	Learn more ☐
Incoming stream *	DerivedTotalSales ⌄	
Filter on *	country!='Unspecified'	

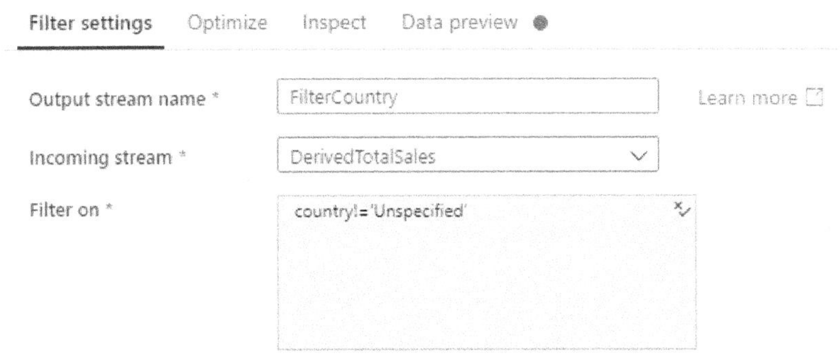

Figure 6.12 – Configuring the filter transformation

19. We'll now add an **Aggregate** transformation to sum the Amount column by country and customer name. To do that, click on the + icon and select the **Aggregate** transformation. Under the **Aggregate settings** tab, change the output stream name to AggregateAmount. Under the **Group by** tab, select the country and customerName columns:

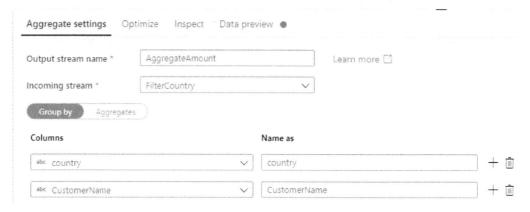

Figure 6.13 – Configuring an aggregate transformation – grouping by columns

20. Switch to the **Aggregates** tab, select **TotalSales**, and enter sum(Amount) in the **Enter expression** box:

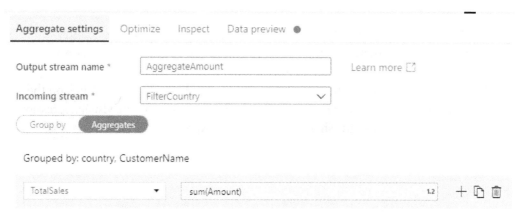

Figure 6.14 – Configuring an aggregate transformation – aggregates

21. The next step is to insert the aggregated data into the Azure SQL database. To do that, click on the + icon and then select **Sink** from the context menu. A new **Sink1** transformation box is added to the data flow canvas. Under the **Sink** tab, name the output stream InsertSalesStaging. From the **Dataset** dropdown, select the **SalesStagingTable** dataset:

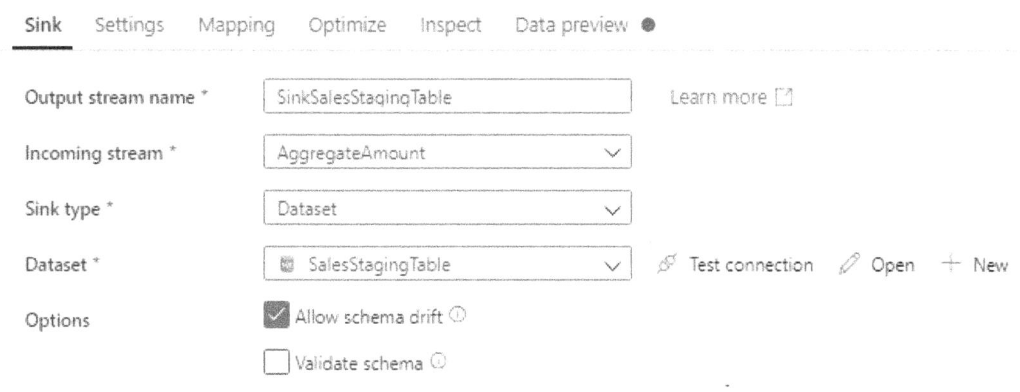

Sink   Settings   Mapping   Optimize   Inspect   Data preview ●

Output stream name *	SinkSalesStagingTable	Learn more ⬁
Incoming stream *	AggregateAmount ⌄	
Sink type *	Dataset ⌄	
Dataset *	⬚ SalesStagingTable ⌄	⌗ Test connection ✐ Open + New
Options	☑ Allow schema drift ⓘ	
	☐ Validate schema ⓘ	

Figure 6.15 – Configuring the sink properties

22. Switch to the **Settings** tab. Select **Truncate table** as the table action. Copy and paste `Execute dbo.MergeSales` in the **Post SQL scripts** text box. The preceding stored procedure will be executed after the data is inserted into the staging table and will upsert the data into the `Sales` table:

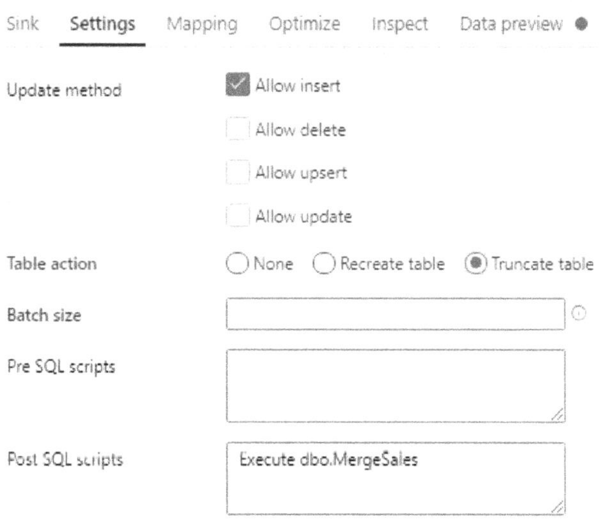

Sink   **Settings**   Mapping   Optimize   Inspect   Data preview ●

Update method	☑ Allow insert
	☐ Allow delete
	☐ Allow upsert
	☐ Allow update
Table action	○ None ○ Recreate table ◉ Truncate table
Batch size	[ ] ⓘ
Pre SQL scripts	[ ]
Post SQL scripts	Execute dbo.MergeSales

Figure 6.16 – Configuring the sink – the Settings tab

23. Switch to the **Mapping** tab. The **Auto mapping** feature is on by default and maps the columns based on name and data type. Turn off the **Auto mapping** option and map the columns as shown in the following screenshot:

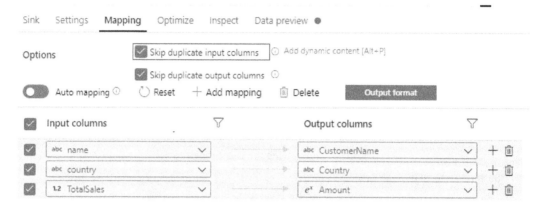

Figure 6.17 – Configuring the sink – mapping

24. Switch to the **Data preview** tab and hit **Refresh**. You should get an output as shown in the following screenshot:

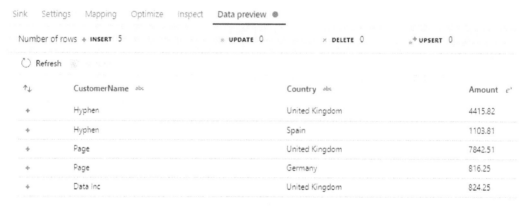

Figure 6.18 – Viewing the sink transformation output

The **Sink** transformation will truncate the `SalesStaging` table, insert the data, and then execute the `dbo.MergeSales` stored procedure to upsert the values into the `Sales` table. Click **Publish all** to save the work.

25. We'll now add another data stream to calculate the running total of the sales and insert it into the `RunningTotal` table in the database. To do this, click on the + icon at the bottom of the **FilterCountry** transformation and select **New branch** from the context menu. A new data stream with **FilterCountry** as the start data stream is added to the canvas. Click on the + icon at the bottom of **FilterCountry** (new branch) and select the **Window** transformation. A new **Window1** transformation box is added to the data flow canvas. Under the **Window settings** tab, name the output stream `Runningtotal`. Leave the **Over** tab (under **Window settings**) as is and switch to the **Sort** tab. For **FilterCountry's column**, select **Amount** and set the sort order to **Ascending**:

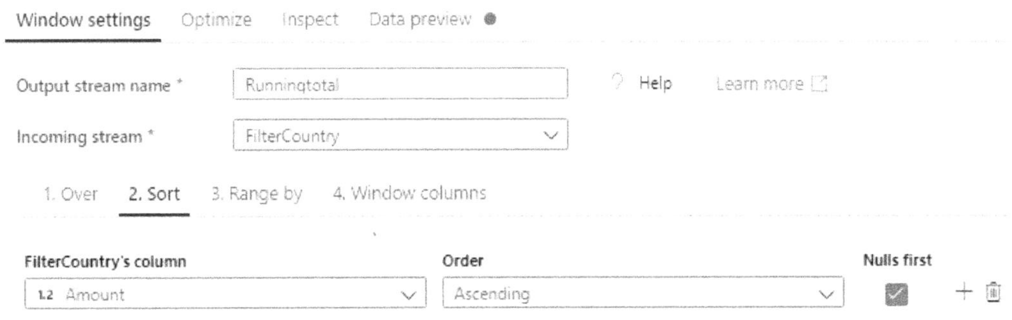

Window settings	Optimize	Inspect	Data preview ●

| Output stream name * | Runningtotal | | ? Help | Learn more ☑ |
| Incoming stream * | FilterCountry ∨ | | | |

1. Over   **2. Sort**   3. Range by   4. Window columns

FilterCountry's column	Order	Nulls first
1.2 Amount ∨	Ascending ∨	☑ + 🗑

Figure 6.19 – Configuring the window transformation – sort

26. Switch to **Window columns**. Click on the **Add or select a column** text box and set the name to `RunningTotal`. Click on the **Enter Expression** text box and copy and paste `sum(Amount)` in the visual expression builder. The **Window** transformation will sort the amount in ascending order and will add the previous row's value to the current row to get the running total:

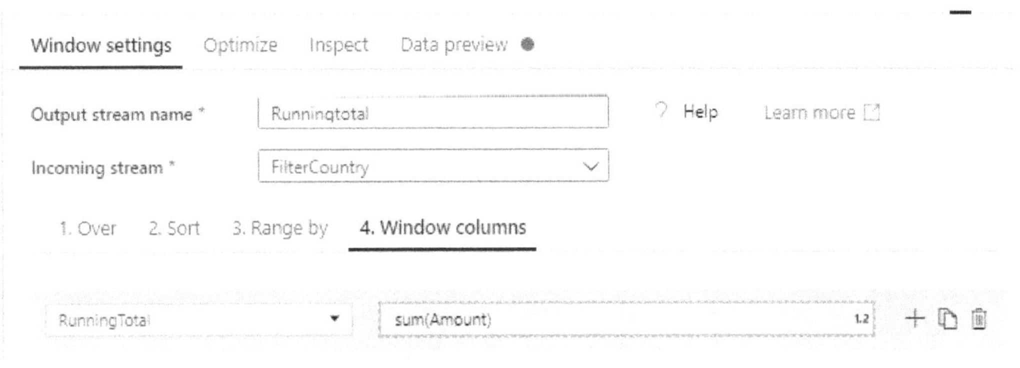

Window settings	Optimize	Inspect	Data preview ●

| Output stream name * | Runningtotal | | ? Help | Learn more ☑ |
| Incoming stream * | FilterCountry ∨ | | | |

1. Over   2. Sort   3. Range by   **4. Window columns**

RunningTotal ▼	sum(Amount)	1.2	+ 🗐 🗑

Figure 6.20 – Configuring the window transformation – Window columns

27. We'll now insert the data from the window transformation into the `RunningTotal` table. To do that, click the + icon at the bottom of the **RunningTotal** transformation and select the **Sink** transformation from the context menu. A new **sink1** transformation box is added to the data flow. Under the **Sink** tab, name the output stream `SinkRunningTotal`. Select the **RunningTotalTable** dataset from the **Dataset** dropdown list:

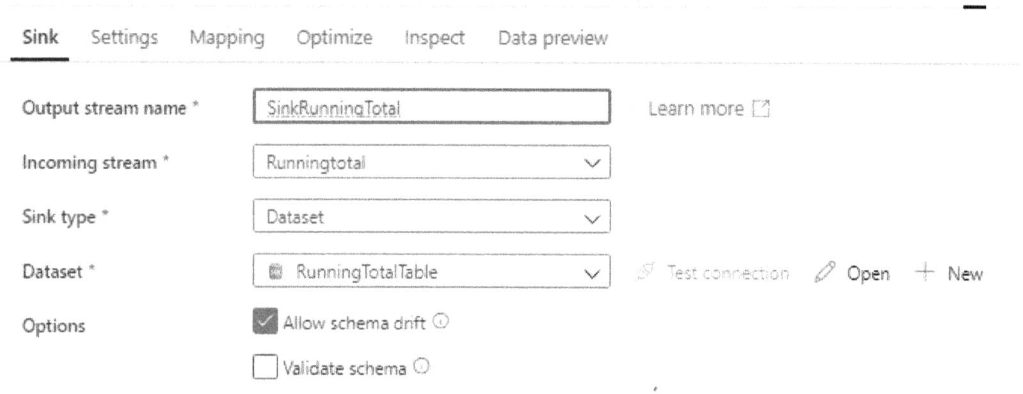

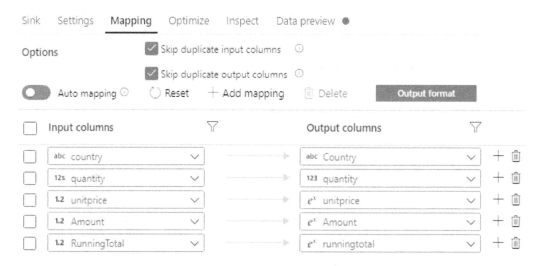

Figure 6.21 – Configuring the running total sink transformation – the Sink tab

28. Switch to the **Mapping** tab. Turn off **Auto mapping** to verify the source and destination column mappings. The mappings should be similar to the following screenshot:

Figure 6.22 – Configuring the sink running total – Mapping

29. This completes the data flow configuration. Click **Publish all** to save your work. The data flow should be similar to the following screenshot:

Figure 6.23 – Mapping data flow

30. Switch to the **Pipeline** tab and select the **Mapping** data flow. Under the **General** tab, name the data flow `Process Orders`. Switch to the **Settings** tab. The **Settings** tab allows us to configure the integration runtime and the compute capacity. The compute capacity determines the data flow performance and the pipeline cost. We can increase or decrease the compute and the core count depending on the amount of data to be processed:

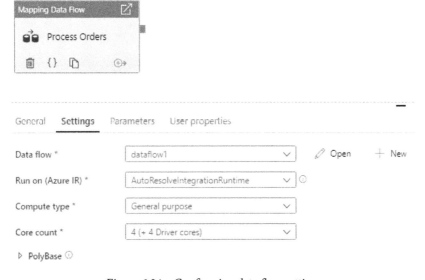

Figure 6.24 – Configuring data flow settings

The minimum available compute and core count is the default option. Leave all the settings as they are.

31. To run the pipeline, click **Debug**. When the pipeline completes, you should get an output as shown in the following screenshot:

Figure 6.25 – Viewing the pipeline output

32. To get the detailed output, click on **Process Orders** (shown in *Figure 8.25*) and click on the spectacles icon. You will get something similar to the following:

Figure 6.26 – Viewing the detailed output

The preceding figure shows detailed output including the time taken and the number of rows processed by each of the data flow components.

33. In the Azure portal, switch to the `packtstorage` account:

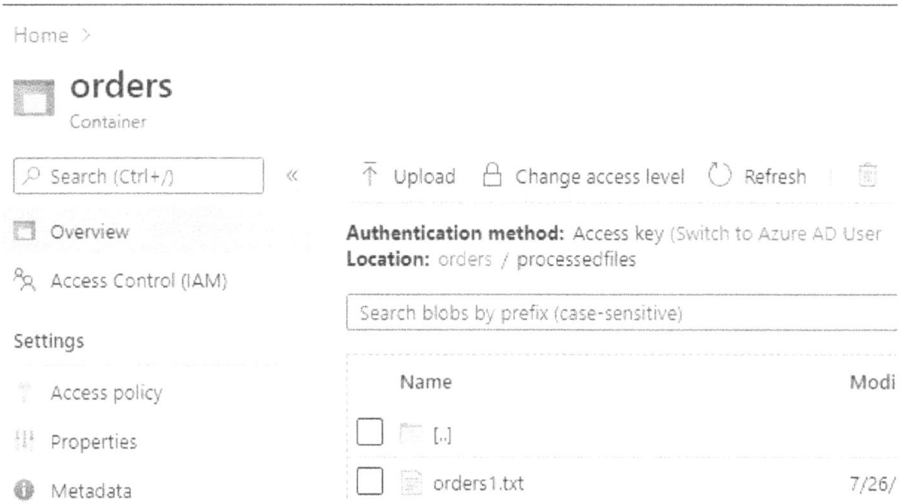

Home >

orders
Container

🔍 Search (Ctrl+/) «	↑ Upload 🔒 Change access level ↻ Refresh 🗑
🔲 Overview	**Authentication method:** Access key (Switch to Azure AD User
ᴿ Access Control (IAM)	**Location:** orders / processedfiles
Settings	Search blobs by prefix (case-sensitive)
☇ Access policy	Name Modi
⊞ Properties	☐ 📁 [..]
ⓘ Metadata	☐ 📄 orders1.txt 7/26/

Figure 6.27 – Viewing the storage account

Observe that as configured, the `orders1.txt` file is moved to the `processfiles` folder from the `datain` folder. We can also query the `Sales` and `RunningTotal` tables in the Azure SQL database to verify whether or not the data is inserted into the tables.

Implementing a wrangling data flow

A wrangling data flow performs code-free data preparation at scale by integrating Power Query to prepare/transform data. The Power Query code is converted to Spark and gets executed on a Spark cluster.

In this recipe, we'll implement a wrangling data flow to read the `orders.txt` file, clean the data, calculate the total sales by country and customer name, and insert the data into an Azure SQL Database table.

Getting ready

To get started, do the following:

1. Log in to `https://portal.azure.com` using your Azure credentials.

2. Open a new PowerShell prompt. Execute the following command to log in to your Azure account from PowerShell:

```
Connect-AzAccount
```

3. You will need an existing Data Factory account. If you don't have one, create one by executing the `~/azure-data-engineering-cookbook\Chapter04\3_ CreatingAzureDataFactory.ps1` PowerShell script.

4. Create an Azure storage account and upload the files to the `~/Chapter06/Data` folder in the `orders/datain` containers. You can use the `~/Chapter06/2_ UploadDatatoAzureStorage.ps1` PowerShell script to create the storage account and upload the files.

5. Follow *steps 1-6* from the previous recipe to create a storage account and upload the `orders.txt` file, create the Azure SQL database and the required database objects, and create the required linked service and the datasets.

How to do it...

Let's start by importing the source dataset schema:

1. The wrangling dataset requires the source dataset schema to be imported. To do that, in the Azure portal, on the **packtdatafactory author** page and under the **Factory Resources** tab, expand **Datasets** and select `orders`:

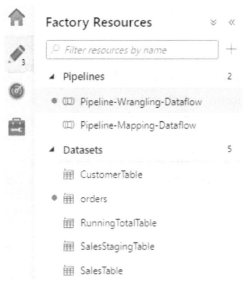

Figure 6.28 – Opening the orders dataset

2. On the **Connection** tab, for **File path**, enter `orders1.txt` as the filename:

Figure 6.29 – Adding the filename

3. On the `orders` dataset tab, click **Import schema** and then select **From connection/store** to import the schema:

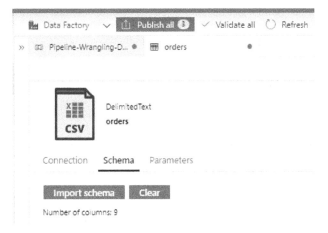

Figure 6.30 – Importing the orders dataset schema

Click **Publish all** to save the changes to the `orders` dataset.

4. Create a new pipeline named `Pipeline-Wrangling-Dataflow`. Drag and drop the **Data flow** activity from the **Move & transform** section. In the **Adding data flow** window, select **Wrangling Data Flow (Preview)** and click **OK**:

Figure 6.31 – Selecting Wrangling Data Flow (Preview)

5. In the **Adding Wrangling Data Flow** window, name the data flow `ProcessOrders` and set the source dataset to `Orders`. Select the sink dataset to **SalesStagingTable**. Select **Truncate table** for the filename option and enter `Execute dbo.MergeSales` as the post SQL scripts:

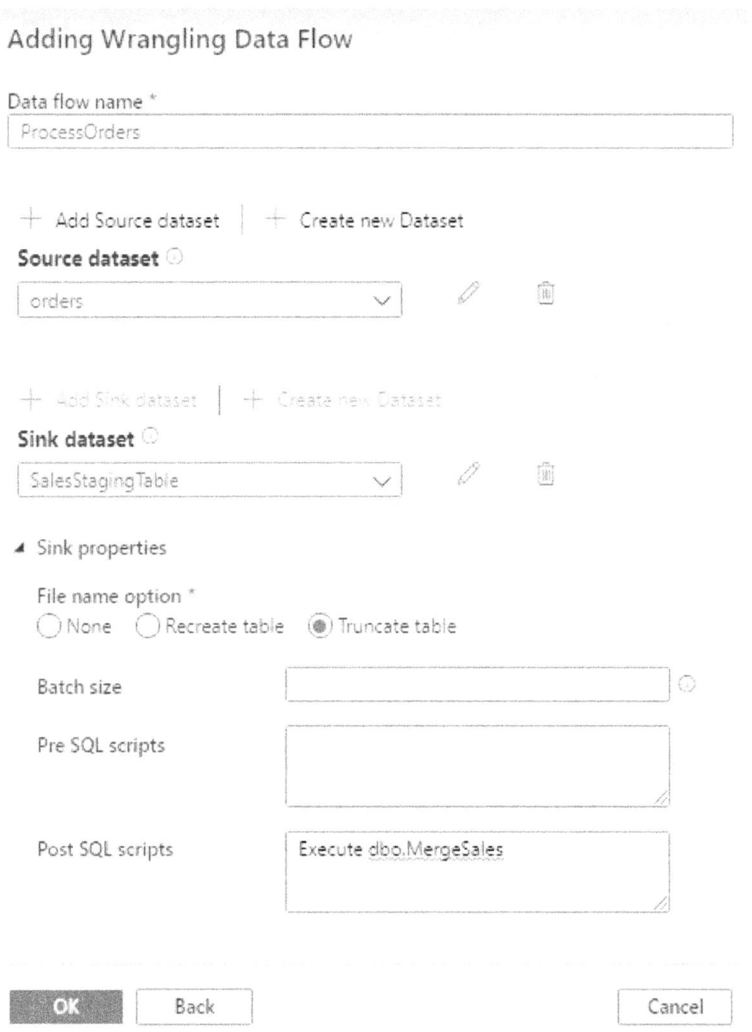

Adding Wrangling Data Flow

Data flow name *

ProcessOrders

+ Add Source dataset | + Create new Dataset

Source dataset ⓘ

orders

+ Add Sink dataset | + Create new Dataset

Sink dataset ⓘ

SalesStagingTable

◢ Sink properties

File name option *
○ None   ○ Recreate table   ● Truncate table

Batch size

Pre SQL scripts

Post SQL scripts   Execute dbo.MergeSales

OK   Back   Cancel

Figure 6.32 – Adding the wrangling data flow

6. Click **OK** to add the data flow. A new wrangling data flow will be added to the pipeline and the Power Query online mashup editor window will open:

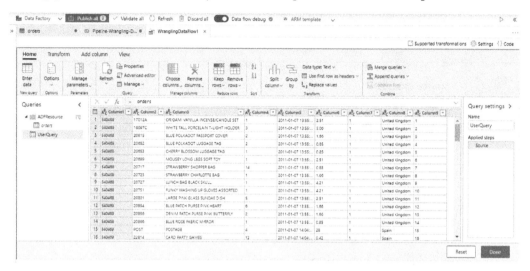

Figure 6.33 – Power Query online mashup editor

The transformation can now be easily done using the Power Query online editor.

7. Let's start by changing the column names. To change a column name, double-click on the column name, say, `Column1`, as shown in *Figure 8.33*, and rename it `invoiceno`. Similarly, change the rest of the column names as follows:

`Column1 = invoiceno`

`Column2 = stockcode`

`Column3 = description`

`Column4 = quantity`

`Column5 = invoicedate`

`Column6 = customerid`

`Column7 = unitprice`

`Column8 = country`

`Column9 = orderid`

8. We'll now change the data type of the `quantity` and `unitprice` columns to integer and decimal. To do that, right-click on the `quantity` column, select **Change type**, and then select **Whole number**:

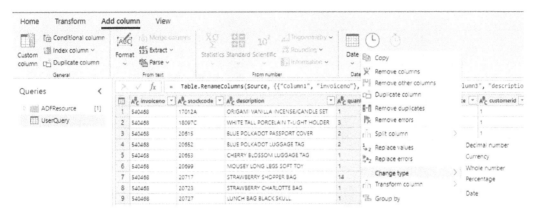

Figure 6.34 – Changing the quantity column's data type

You can also change the data type by selecting the **ABC** icon next to the `quantity` column and then selecting **Whole number**. Similarly, change the data type of the `unitprice` column to **Decimal number** and the `customerid` column to **Whole number**.

9. We'll now add a new column, `Amount`. The `Amount` column is the product of the `quantity` and `unitprice` columns. To do that, click the **Add column** menu option and then select **Custom column**:

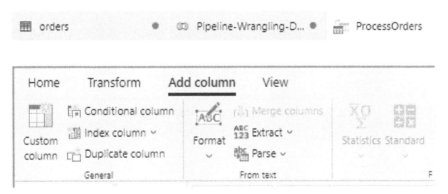

Figure 6.35 – Adding a new custom column

10. In the **Custom column** window, enter the column name as `Amount` and the custom column formula as `[quantity]*[unitprice]`:

Figure 6.36 – Configuring the custom column

11. We need to add customer names to our dataset. To do that, we'll need to join the `Customer` table in Azure SQL Database with the `Orders` dataset. Click on **Settings** at the top right of the data flow window:

Figure 6.37 – Adding a new data source

12. The **Adding Wrangling Data Flow** window will open. Click **Add Source dataset** and select **CustomerTable**:

Figure 6.38 – Adding the CustomerTable source dataset

13. Click **OK** to save and go back to the Power Query mashup editor window. Change the data type of the Amount column to **Decimal number**.

14. To join the orders and customertable datasets, select **Merge queries** from the **Home** menu:

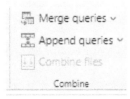

Figure 6.39 – Merging datasets

15. In the **Merge** window, under **UserQuery**, select the `customerid` column. Select **CustomerTable** from the **Right table for merge** option and then select the **id** column. Select **Inner** as the join kind. The two `Customertable` and `UserQuery` queries will be joined on the `UserQuery.customerid` and `CustomerTable.id` columns:

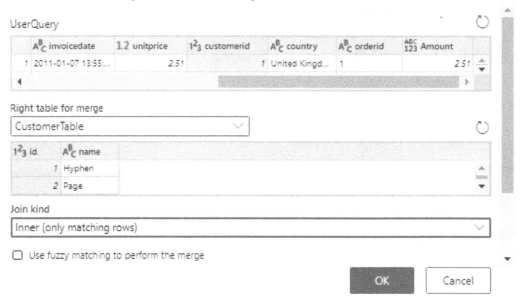

Figure 6.40 – Merging/joining UserQuery and CustomerTable datasets

16. Click **OK** to apply the join. A new **CustomerTable** column is added to `UserQuery`. Click on the expand icon next to the column. In the expanded window, uncheck the `id` column as it's not required:

Figure 6.41 – Expanding CustomerTable

17. Click **OK** to add the name column to the user query. Rename the
    `CustomerTable.name` column `CustomerName`.

18. To aggregate the amount for country and customer name, right-click the `Amount`
    column and select **Group by** from the context menu. In the **Group by** window,
    select **Advanced**. Click the **Add grouping** button. Add country and customer name
    as the grouping columns. Enter `TotalSales` as the new column name, select **Sum**
    as the operation, and set **Amount** as the column:

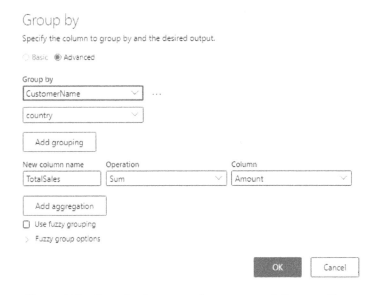

Figure 6.42 – Grouping by amount for country and CustomerName

Click **OK** to continue. You should get an output as shown in the following screenshot:

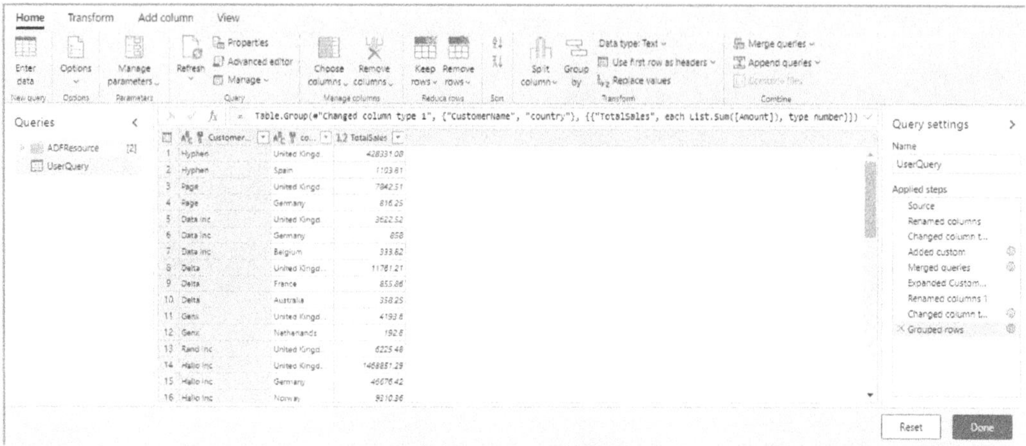

Figure 6.43 – Grouping output

19. Click **Done** and then click **Publish all** to save the work. This completes the data flow configuration.

20. To run the pipeline, click **Debug**. Make sure that **Data flow debug** mode is on. When the pipeline completes, you'll get an output as shown in the following screenshot:

Parameters   Variables   Output

Pipeline run ID: 220a566f-1dc7-49ce-adb4-79324172bff5 [@]   ⟳   ◯

NAME	TYPE	RUN START	DURATION	STATUS
ProcessOrders ⇥ ⇤ ⟲	ExecuteWrang	2020-07-26T07:55:07.708	00:01:33	✅ Succeeded

Figure 6.44 – Viewing pipeline output

21. Click on the spectacles icon next to **ProcessOrder** to view the detailed output:

Figure 6.45 – Viewing detailed output

The detailed output shows the time taken and the rows against each of the transformations. The data flow ran 12 transformations in under 2 minutes to process the data.

7
Azure Data Factory Integration Runtime

The Azure Data Factory **Integration Runtime** (**IR**) is the compute infrastructure that is responsible for executing data flows, pipeline activities, data movement, and **SQL Server Integration Services** (**SSIS**) packages. There are three types of IR: Azure, self-hosted, and Azure SSIS.

The Azure IR is the default IR that is created whenever a new data factory is created. It can process data flows, data movement, and activities.

A self-hosted IR can be installed on-premises or on a virtual machine running the Windows OS. A self-hosted IR can be used to work with data on-premises or in the cloud. It can be used for data movement and activities.

The Azure SSIS IR is used to lift and shift existing SQL SSIS.

In this chapter, we'll learn how to use a self-hosted IR and Azure SSIS IR through the following recipes:

- Configuring a self-hosted IR
- Configuring a shared self-hosted IR
- Migrating an SSIS package to Azure Data Factory
- Executing an SSIS package with an on-premises data store

> **Note**
> The wrangling data flow is in preview at the time of writing this book.

Technical requirements

For this chapter, the following are required:

- A Microsoft Azure subscription
- PowerShell 7
- Microsoft Azure PowerShell

Configuring a self-hosted IR

In this recipe, we'll learn how to configure a self-hosted IR and then use the IR to copy files from on-premises to Azure Storage using the **Copy data** activity.

Getting ready

To get started, do the following:

1. Log in to `https://portal.azure.com` using your Azure credentials.

2. Open a new PowerShell prompt. Execute the following command to log in to your Azure account from PowerShell:

   ```
   Connect-AzAccount
   ```

3. You will need an existing Data Factory account. If you don't have one, create one by executing the `~/azure-data-engineering-cookbook\Chapter04\3_CreatingAzureDataFactory.ps1` PowerShell script.

How to do it...

To configure a self-hosted runtime, follow the given steps:

1. In the Azure portal, open Data Factory, and then open the **Manage** tab:

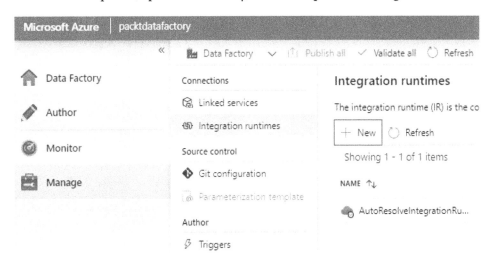

Figure 7.1 – Opening the Manage tab – Data Factory

2. Select **New**, and then select **Azure, Self-Hosted**:

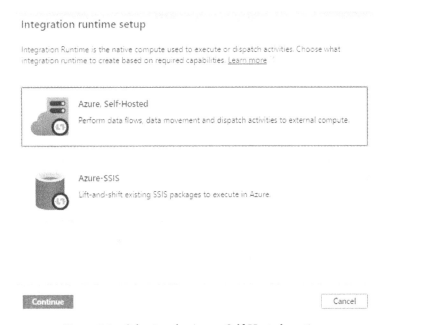

Figure 7.2 – Selecting the Azure, Self-Hosted runtime

3. Click **Continue** to go to the next step. In the **Network environment** section, select **Self-Hosted**:

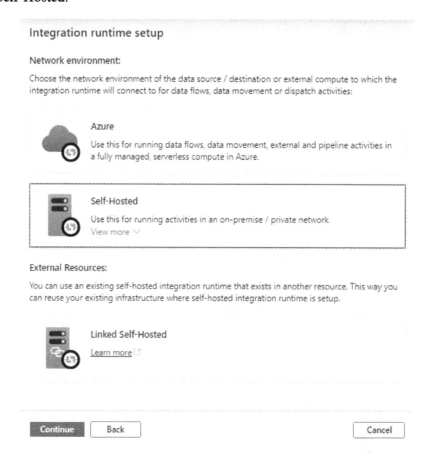

Figure 7.3 – Selecting a network environment

4. Click **Continue** to go to the next step. In the next window, name the IR selfhosted-onpremise and click **Create**:

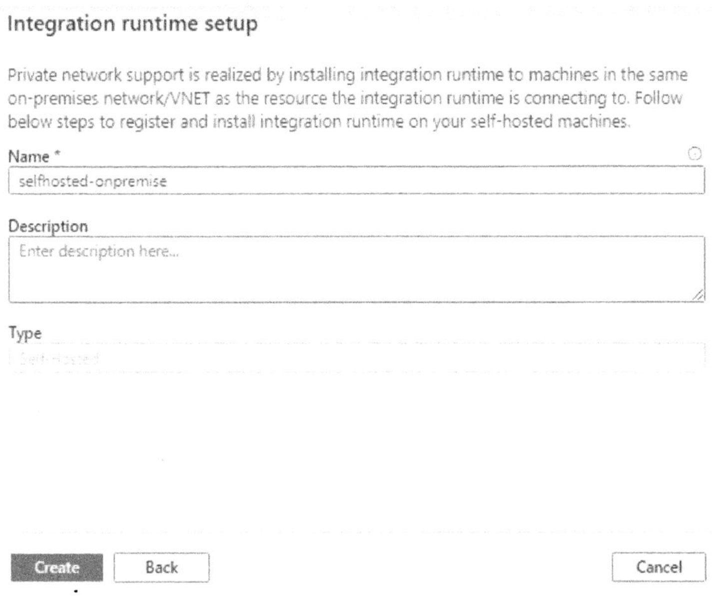

Integration runtime setup

Private network support is realized by installing integration runtime to machines in the same
on-premises network/VNET as the resource the integration runtime is connecting to. Follow
below steps to register and install integration runtime on your self-hosted machines.

Name *

selfhosted-onpremise

Description

Enter description here...

Type

Self-Hosted

Create   Back   Cancel

Figure 7.4 – Creating the IR

5. The next step is to download and install the runtime on the on-premises machine:

Integration runtime setup

Settings   Nodes   Auto update   Sharing

Install integration runtime on Windows machine or add further nodes using the
Authentication Key.

Name

selfhosted-onpremise

Option 1: Express setup

Click here to launch the express setup for this computer

Option 2: Manual setup

Step 1: Download and install integration runtime

Step 2: Use this key to register your integration runtime

NAME	AUTHENTICATION KEY
Key1	IR@f1220976-49a2-4595-af44-45224e7cdbbf@packtdatafactory@cu@7t
Key2	IR@f1220976-49a2-4595-af44-45224e7cdbbf@packtdatafactory@cu@ih

Close

Figure 7.5 – Installing the IR

Note down the **Key1** value as it's required in a later step.

If you are logged in to the Azure portal from the machine where you wish to install the runtime, select **Option 1, Express setup**. If you want to install it onto another machine, select **Option 2, Download and install integration runtime**.

6. We'll be using **Option 2**. Click on **Download and install integration runtime**. Download the latest version from the download center. Double-click the downloaded file and click **Run** to start the installation wizard. Follow the wizard to install the runtime. When the installation completes, copy and paste the authentication key and click **Register**:

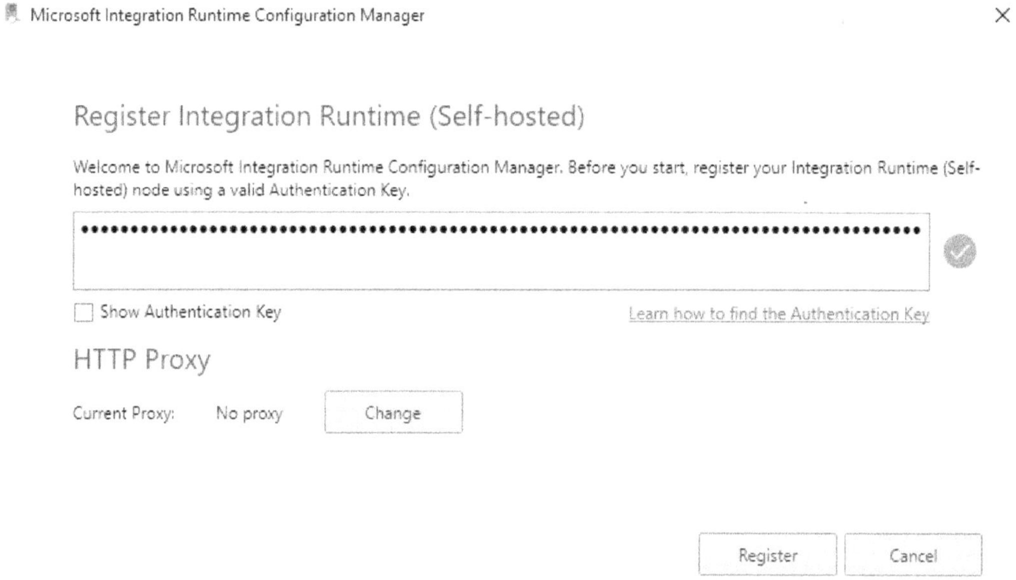

Figure 7.6 – Registering a self-hosted IR

7. After the successful verification of the key, the computer or the node is registered as part of the `selfhosted-onpremise` IR:

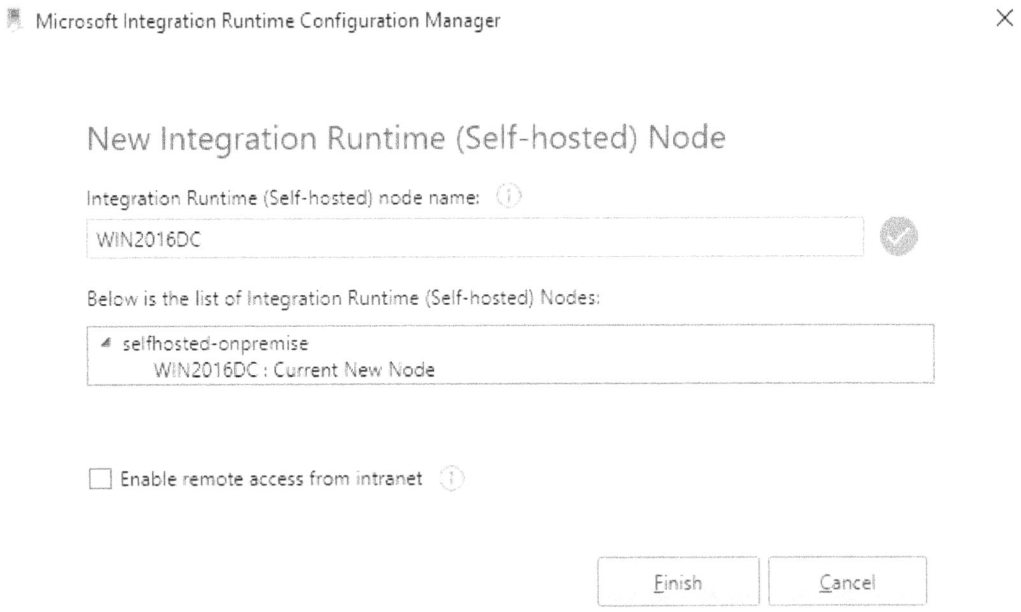

Figure 7.7 – Completing self-hosted IR node configuration

8. Switch to the Azure portal and close the IR setup window. The new `selfhosted-onpremise` IR will be listed in the **Integration runtimes** window:

Integration runtimes

Time zone : Chennai, Kolkata, Mumbai, New...

All   Self-Hosted   Azure   Azure-SSIS   ○ Refresh   ≡≡ Edit columns

Showing 1 - 2 of 2 items

NAME ↑↓	TYPE ↑↓	SUB-TYPE ↑↓	STATUS ↑↓	REGION ↑↓
AutoResolveIntegrationRuntime	Azure	Public	✓ Running	Auto Resolve
selfhosted-onpremise	Self-Hosted	---	✓ Running	---

Figure 7.8 – Viewing the self-hosted IR status

9.   Click on the **selfhosted-onpremise** IR to view the details:

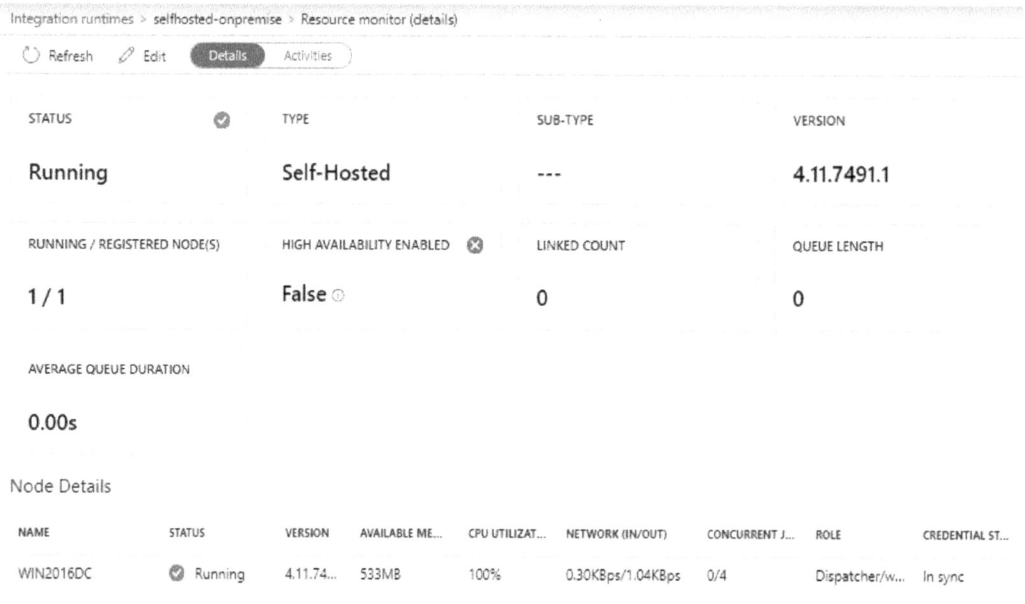

Integration runtimes > selfhosted-onpremise > Resource monitor (details)

◯ Refresh   ✎ Edit   ▐ Details ▌   Activities

STATUS		TYPE	SUB-TYPE	VERSION
Running	✔	Self-Hosted	---	4.11.7491.1

RUNNING / REGISTERED NODE(S)	HIGH AVAILABILITY ENABLED	LINKED COUNT	QUEUE LENGTH
1 / 1	False ⊘	0	0

AVERAGE QUEUE DURATION

0.00s

Node Details

NAME	STATUS	VERSION	AVAILABLE ME...	CPU UTILIZAT...	NETWORK (IN/OUT)	CONCURRENT J...	ROLE	CREDENTIAL ST...
WIN2016DC	✔ Running	4.11.74...	533MB	100%	0.30KBps/1.04KBps	0/4	Dispatcher/w...	In sync

Figure 7.9 – Viewing the self-hosted IR details

The WIN2016DC node is the name of the computer on which we installed the runtime. We can add more nodes by installing the runtime and then authenticating with the IR key.

We'll now use the self-hosted IR to upload data from the files stored on-premises to Azure SQL Database.

Note

To troubleshoot a self-hosted IR on a machine, open the IR on the local machine and view the logs in the diagnostic tab. You can open the self-hosted IR by clicking on the server icon in the system tray.

---

10. We'll now create a linked service to connect to the files in the on-premises system. To do that, open ~/Chapter07/ OnpremiseFileLinkedServiceDefinition.txt in a notepad:

```
{
    "name": "Onpremisefilelinkedservice",
    "type": "Microsoft.DataFactory/factories/
linkedservices",
```

```
    "properties": {
      "annotations": [],
      "type": "FileServer",
      "typeProperties": {
        "host": "C:\\azure-data-engineering-cookbook\\
  Chapter07\\Data",
          "userId": "WIN2016DC\\administrator",
          "password": {
          "type":"SecureString",
          "value":"Awesome@1234"
      },
      "connectVia": {
        "referenceName": "selfhosted-onpremise",
        "type": "IntegrationRuntimeReference"
      }
    }
  }
```

Modify the host value to the path that contains the files to be uploaded to Azure SQL Database. Provide the local system user ID and password that have access to read the files. In our case, we have provided the local machine administrator user. Make sure that the user has either the machine or domain name in the prefix.

Observe that the runtime specified under connectVia is the selfhosted-onpremise IR created earlier.

11. Azure Data Factory connects to the local computer through the self-hosted runtime. Save and close the file. Execute the following PowerShell command to create the linked service:

```
Set-AzDataFactoryV2LinkedService -Name
Onpremisefilelinkedservice -DefinitionFile .\
azure-data-engineering-cookbook\Chapter07\
OnpremiseFileLinkedServiceDefinition.txt
-ResourceGroupName packtade -DataFactoryName
packtdatafactory
```

The preceding command creates Onpremisefilelinkedservice in packtdatafactory.

12. We'll now create the dataset for the files in the ~/Chapter07/Data folder. To do that, open ~/Chapter07/OrdersDatasetDefinition.txt and change the linked service name under linkedServiceName| referenceName if required. Save and close the file. Execute the following PowerShell command to create the ordersonpremise dataset:

```
Set-AzDataFactoryV2LinkedService -Name ordersonpremise
-DefinitionFile .\azure-data-engineering-cookbook\
Chapter07\OrdersonpremiseDatasetDefinition.
txt -ResourceGroupName packtade -DataFactoryName
packtdatafactory
```

The preceding command creates the ordersonpremise dataset for the files hosted locally.

13. The next step is to create an Azure storage account to which the files on the local system will be copied. To do that, execute the following PowerShell command:

```
.\azure-data-engineering-cookbook\Chapter06\2_
UploadDatatoAzureStorage.ps1 -resourcegroupname packtade
-storageaccountname packtstorage -location centralus -
createstorageaccount $true -uploadfiles $false
```

The preceding command creates a storage account called packtstorage and a container called orders. Copy and save the storage account key from the command output to be used in later steps.

14. We'll now create the linked service for the storage account. To do that, open ~/Chapter07/PacktStorageLinkedService.txt in a notepad. In the connectionString parameter, modify the account name if required and enter the key copied in the previous step Save and close the file. Execute the following command to create the linked service:

```
Set-AzDataFactoryV2LinkedService -Name
PacktStorageLinkedService -DefinitionFile .\
azure-data-engineering-cookbook\Chapter07\
PacktStorageLinkedServiceDefinition.txt
-ResourceGroupName packtade -DataFactoryName
packtdatafactory
```

15. We'll now create the dataset for the Azure storage account. To do that, open ~/ Chapter07/OrdersDatasetDefinitionFile.txt. Modify the linked service name if required. Save and close the file. Execute the following PowerShell command to create the dataset:

```
New-AzDataFactoryV2Dataset -Name ordersazure
-DefinitionFile .\azure-data-engineering-cookbook\
Chapter07\OrdersDatasetDefinition.txt -ResourceGroupName
packtade -DataFactoryName packtdatafactory
```

The preceding command creates the ordersazure dataset with the location as / orders/datain.

16. We'll now configure the copy data activity to copy the data from the ordersonpremise dataset to the ordersazure dataset. To do that, open the packtdatafactory author page and create a new pipeline. Name the pipeline Pipeline-Selfhosted-IR. Under the **Activities** tab, drag and drop the **Copy data** activity from the **Move & transform** section. Under the **General** tab, name the copy data activity Copy data-on-premise to Azure Storage. Under the **Source** tab, select the ordersonpremise dataset from the **Source dataset** dropdown. Under **File path type**, select **Wildcard file path** and enter *orders* as the wildcard. This will copy only the files with the word orders in the filename:

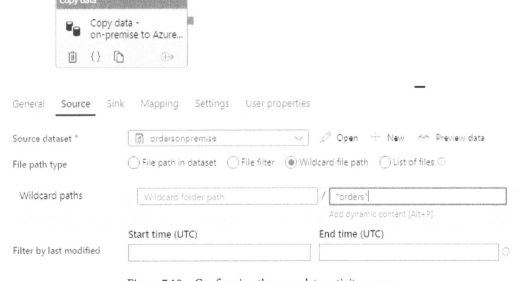

Figure 7.10 – Configuring the copy data activity source

17.   Under the **Sink** tab, from the **Sink dataset** dropdown, select `ordersazure`:

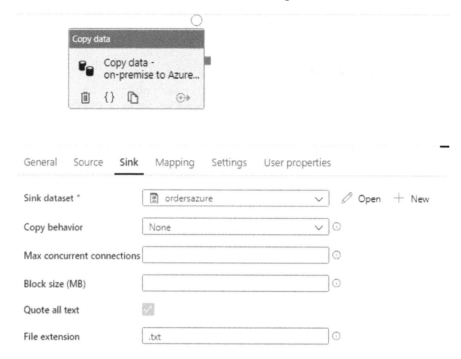

Figure 7.11 – Configuring the copy data activity sink

This completes the configuration. Click **Publish all** to save your work.

18. Click **Debug** to run the pipeline. You should get an output as shown in the following screenshot when the pipeline starts:

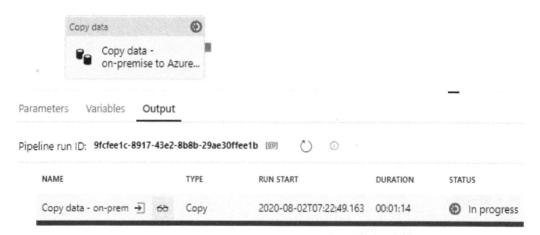

Figure 7.12 – Monitoring the pipeline progress

19. Click on the spectacles icon beside the copy data activity to monitor the runtime progress:

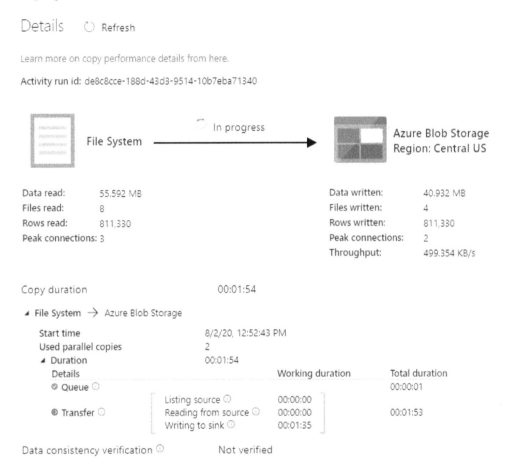

Figure 7.13 – Monitoring real-time execution progress

Configuring a shared self-hosted IR

A shared self-hosted runtime, as the name suggests, can be shared among more than one data factory. This helps to use a single self-hosted IR to run multiple pipelines. In this activity, we'll learn how to share a self-hosted IR.

Getting ready

To get started, do the following:

1. Log in to `https://portal.azure.com` using your Azure credentials.

2. Open a new PowerShell prompt. Execute the following command to log in to your Azure account from PowerShell:

    ```
    Connect-AzAccount
    ```

3. You will need an existing Data Factory account. If you don't have one, create one by executing the `~/azure-data-engineering-cookbook\Chapter04\3_CreatingAzureDataFactory.ps1` PowerShell script.

4. You need a self-hosted IR. If you don't have one, follow the previous recipe to create one.

How to do it...

Let's start by creating a new Azure data factory:

1. Execute the following PowerShell command to create an Azure data factory:

    ```
    .\azure-data-engineering-cookbook\Chapter04\3_
    CreatingAzureDataFactory.ps1 -resourcegroupname packtade
    -location centralus -datafactoryname packtdatafactory2
    ```

    The preceding command used the `Set-AzDataFactoryV2` command to provision a new Azure data factory, `packtdatafactory2`.

2. In the Azure portal, open `packtdatafactory`, and then open the **Manage** tab. In the **Manage** tab, select **Integration runtimes**:

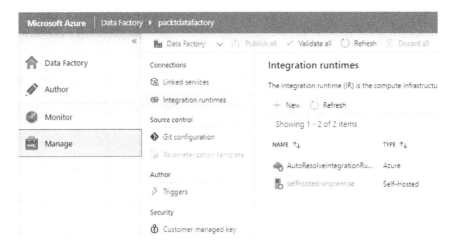

Figure 7.14 – Opening the packtdatafactory Manage tab

Select the `selfhosted-onpremise` IR.

Note

This recipe requires an existing self-hosted IR. If you don't have one, follow the previous recipe to create one.

---

3. In the **Edit integration runtime** window, switch to the **Sharing** tab. Click on **Grant permission to another Data Factory**:

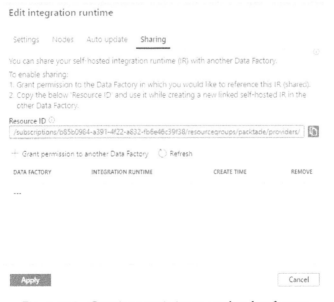

Figure 7.15 – Granting permission to another data factory

4. In the **Assign permissions** window, type `packtdatafactory2` in the search box, and then select **packtdatafactory2** from the list:

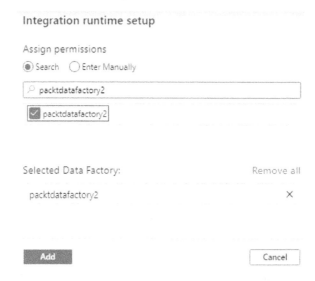

Figure 7.16 – Integration runtime setup window

5. Click **Add** to continue. The data factory is added to the list:

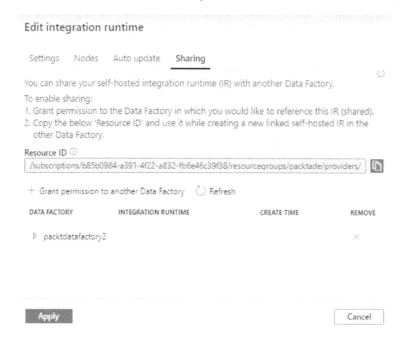

Figure 7.17 – Granting permission to another data factory

Note the resource ID for use in later steps. Click **Apply** to save the configuration settings.

6. In the Azure portal, open the `packtdatafactory2` **Manage** tab. Under the **Connections** tab, select **Integration runtimes**. In the **Integration runtimes** panel, click **+ New**:

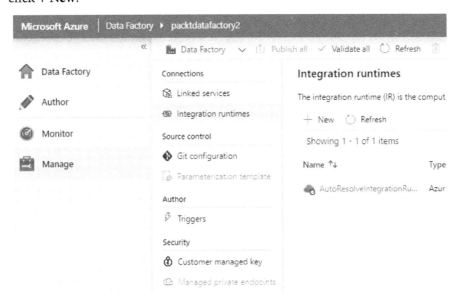

Figure 7.18 – Adding the IR to packtdatafactory2

7. In the **Integration runtime setup** window, select **Azure, Self-Hosted**:

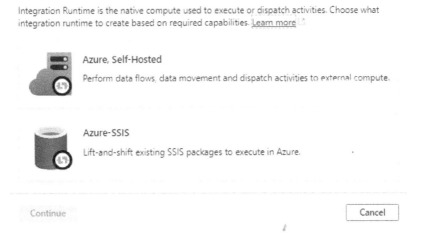

Figure 7.19 – Selecting the IR type

Click **Continue** to go to the next step.

8.   In the next window, under **External Resources**, select **Linked Self-Hosted**:

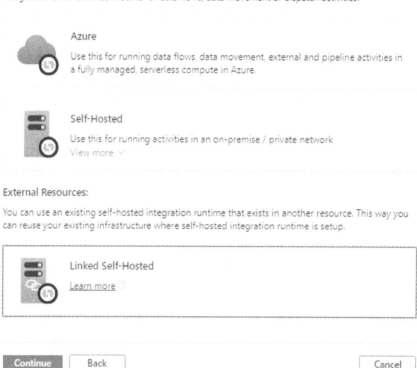

Figure 7.20 – Selecting Linked Self-Hosted

Click **Continue** to go to the next step.

9. In the next window, provide the name as `selfhosted-onpremise`. Under **Resource ID**, enter the **Resource ID** value from *step 3*:

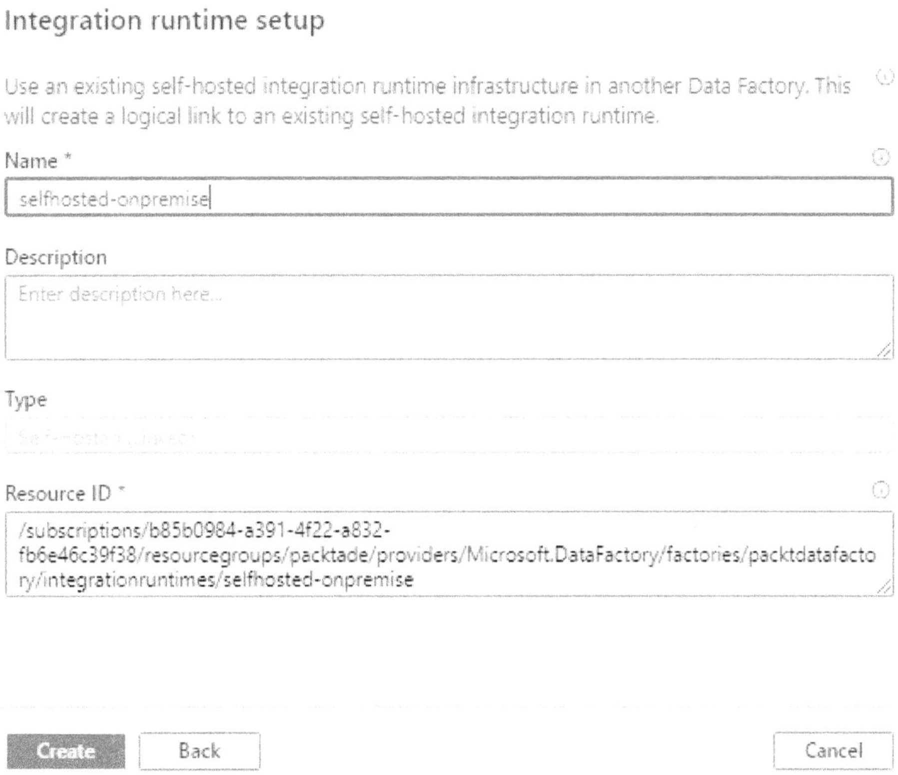

Integration runtime setup

Use an existing self-hosted integration runtime infrastructure in another Data Factory. This
will create a logical link to an existing self-hosted integration runtime.

Name *

> selfhosted-onpremise

Description

> Enter description here...

Type

> Self-hosted (Linked)

Resource ID *

> /subscriptions/b85b0984-a391-4f22-a832-
> fb6e46c39f38/resourcegroups/packtade/providers/Microsoft.DataFactory/factories/packtdatafacto
> ry/integrationruntimes/selfhosted-onpremise

Create   Back   Cancel

Figure 7.21 – Providing a shared IR resource ID

The IR is added to `packtdatafactory2`. The `selfhosted-onpremise` runtime is now shared between `packtdatafactory` and `packtdatafactory2`. The pipelines from the two data factories can benefit from one self-hosted runtime.

Migrating an SSIS package to Azure Data Factory

SQL SSIS is a widely used on-premises ETL tool. In this recipe, we'll learn how to migrate an existing SSIS package to Azure Data Factory.

We'll do this by configuring an Azure SSIS IR, uploading the SSIS package to Azure SQL Database SSISDB, and then executing the package using the Execute SSIS Package activity.

Getting ready

To get started, do the following:

1. Log in to `https://portal.azure.com` using your Azure credentials.

2. Open a new PowerShell prompt. Execute the following command to log in to your Azure account from PowerShell:

    ```
    Connect-AzAccount
    ```

3. You will need an existing Data Factory account. If you don't have one, create one by executing the `~/azure-data-engineering-cookbook\Chapter04\3_CreatingAzureDataFactory.ps1` PowerShell script.

4. Provision an Azure storage account and upload files to it using `~/Chapter06/2_UploadDatatoAzureStorage.ps1`.

How to do it...

We'll start by creating a new Azure SSIS IR:

1. In the Azure portal, open the Data Factory **Manage** tab. Select **Integration runtimes**, and then select **+ New**. Provide the name, location, node size, and node number. We have kept the node size and node number to the smallest available to save costs:

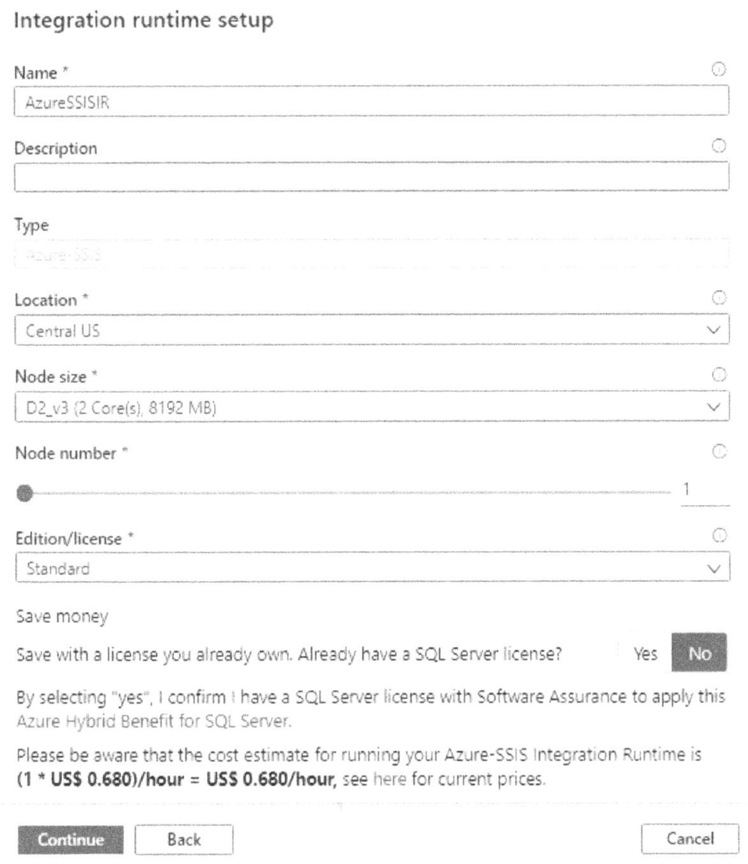

Integration runtime setup

Name *

AzureSSISIR

Description

Type

Azure-SSIS

Location *

Central US

Node size *

D2_v3 (2 Core(s), 8192 MB)

Node number *

1

Edition/license *

Standard

Save money

Save with a license you already own. Already have a SQL Server license?   Yes   No

By selecting "yes", I confirm I have a SQL Server license with Software Assurance to apply this Azure Hybrid Benefit for SQL Server.

Please be aware that the cost estimate for running your Azure-SSIS Integration Runtime is **(1 * US$ 0.680)/hour = US$ 0.680/hour,** see here for current prices.

Continue   Back   Cancel

Figure 7.22 – Configuring the Azure SSIS IR

Click **Continue** to go to the next step.

We can either host the SSIS package in an SSISDB database hosted either on Azure SQL Database or SQL Managed Instance, or we can host the package on Azure Storage. In this recipe, we'll host the package on SSISDB.

2. In the **Deployment settings** section, check the **Create SSIS catalog** box and provide
    the Azure SQL Database details, as shown in the following screenshot. We have
    selected the Azure SQL Database **Basic** performance tier to save on costs:

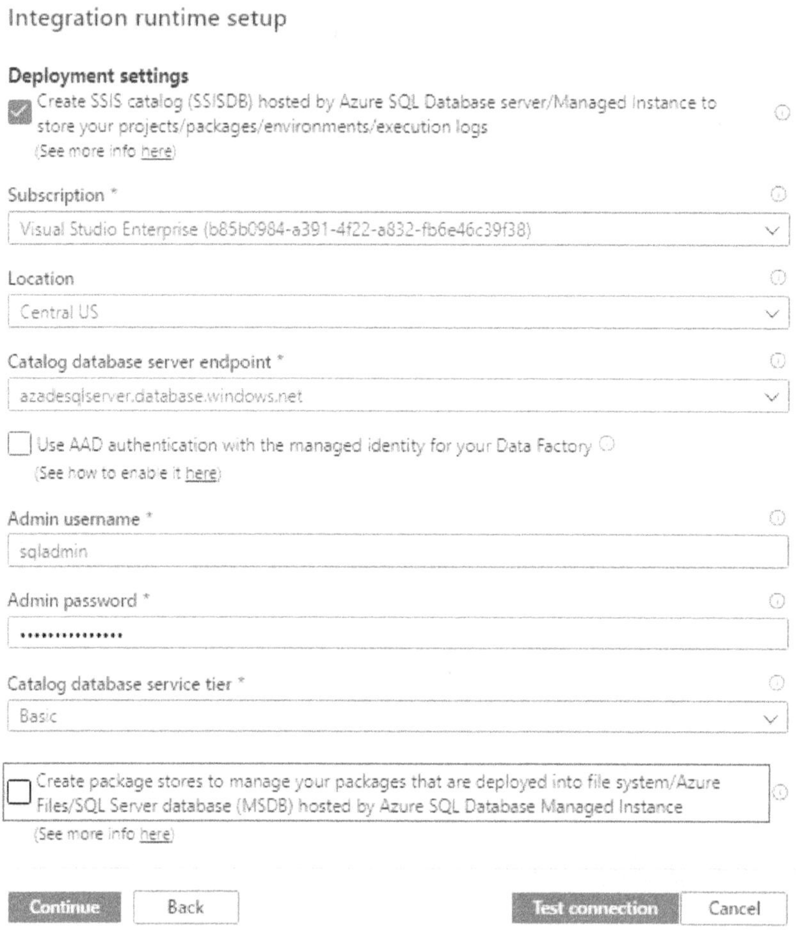

Integration runtime setup

Deployment settings

☑ Create SSIS catalog (SSISDB) hosted by Azure SQL Database server/Managed Instance to   ⓘ
  store your projects/packages/environments/execution logs
  (See more info here)

Subscription *   ⓘ

Visual Studio Enterprise (b85b0984-a391-4f22-a832-fb6e46c39f38)   ⌄

Location   ⓘ

Central US   ⌄

Catalog database server endpoint *   ⓘ

azadesqlserver.database.windows.net   ⌄

☐ Use AAD authentication with the managed identity for your Data Factory ◯
  (See how to enable it here)

Admin username *   ⓘ

sqladmin

Admin password *   ⓘ

••••••••••••••

Catalog database service tier *   ⓘ

Basic   ⌄

☐ Create package stores to manage your packages that are deployed into file system/Azure   ⓘ
  Files/SQL Server database (MSDB) hosted by Azure SQL Database Managed Instance
  (See more info here)

[Continue]   [Back]   [Test connection]   [Cancel]

Figure 7.23 – Configuring the SSISDB catalog

Click **Continue** to go to the next step.

3. On the **Advanced settings** page, leave the settings as is (default) and click **Continue** to create the Azure SSIS IR:

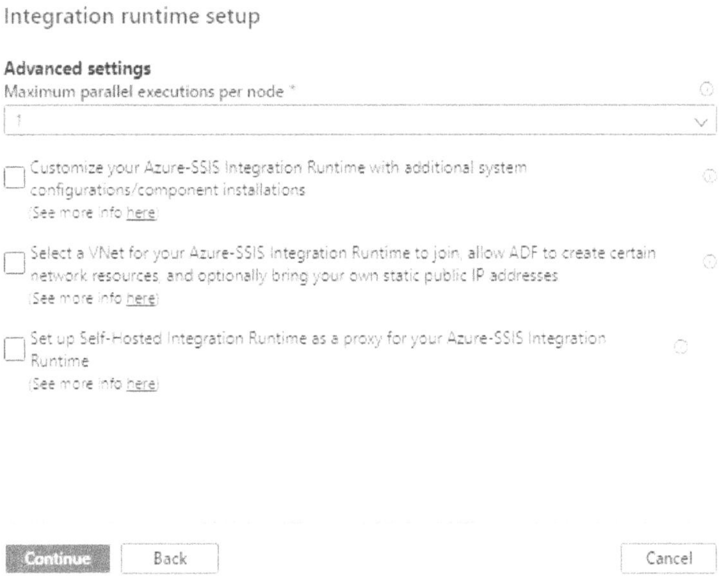
Integration runtime setup
Advanced settings
Maximum parallel executions per node *
1
Customize your Azure-SSIS Integration Runtime with additional system configurations/component installations
(See more info here)
Select a VNet for your Azure-SSIS Integration Runtime to join, allow ADF to create certain network resources, and optionally bring your own static public IP addresses
(See more info here)
Set up Self-Hosted Integration Runtime as a proxy for your Azure-SSIS Integration Runtime
(See more info here)
Continue
Back
Cancel

Figure 7.24 – Configuring advanced settings

When created, the `AzureSSISIR` IR is listed as in the following screenshot:

Microsoft Azure | packtdatafactory
Data Factory
Publish all
Validate all
Refresh
Discard all
Data flow debug
ARM template
Connections
Linked services
Integration runtimes
Source control
Git configuration
Author
Triggers
Security
Integration runtimes
New
Refresh
Showing 1 - 3 of 3 items
NAME
TYPE
SUB-TYPE
STATUS
AzureSSISIR
Azure-SSIS
Running
Public
Running
selfhosted-onpremise
Self-Hosted
Shared
Running

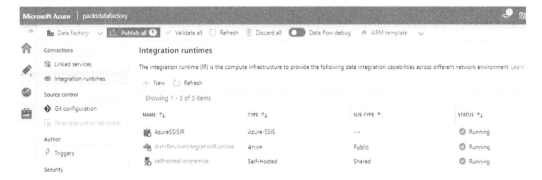

Figure 7.25 – Viewing AzureSSISIR

4. The next step is to deploy the SSIS package to the Azure SQL Database SSISDB catalog. To do that, open SQL Server Management Studio and connect to the `SSISDB` Azure SQL database in the Object Explorer. In the **Object Explorer** window, expand **Integration Services Catalogs**, right-click on **SSISDB**, and then select **Create Folder…**:

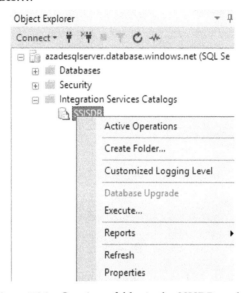

Figure 7.26 – Creating a folder in the SSISDB catalog

5. In the **Create Folder** dialog box, set **Folder name** to `AzureSSIS`:

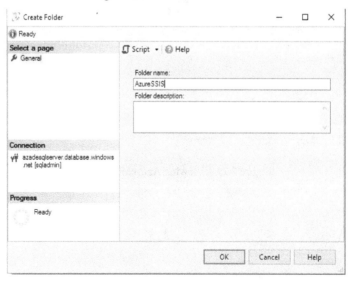

Figure 7.27 – Providing the folder settings

Click **OK** to create the folder.

6. In the **Object Explorer** window, expand the `AzureSSIS` folder, right-click on **Projects**, and select **Deploy Project...** from the context menu:

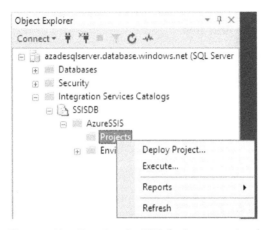

Figure 7.28 – Opening the SSIS deployment wizard

7. In the **Integration Services Deployment Wizard** window, select the **Source** tab, browse to the `~\azure-data-engineering-cookbook\Chapter07\ CopyFiles\CopyFiles\bin\Development` path, and select the `CopyFiles.ispac` file:

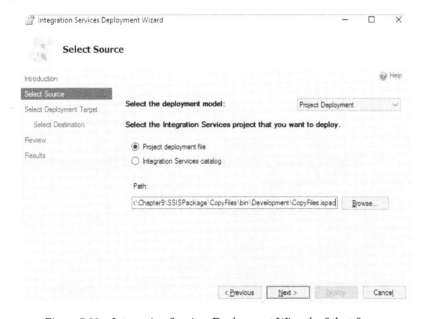

Figure 7.29 – Integration Services Deployment Wizard – Select Source

> **Note**
>
> The `CopyFiles.dtsx` SSIS package copies the files from an Azure storage account from the `orders/datain` folder to the `orders/dataout` folder. The Azure storage account name and key are passed as parameters.

Click **Next** to go to the next step.

8. In the **Select Deployment Target** tab, select **SSIS in Azure Data Factory**:

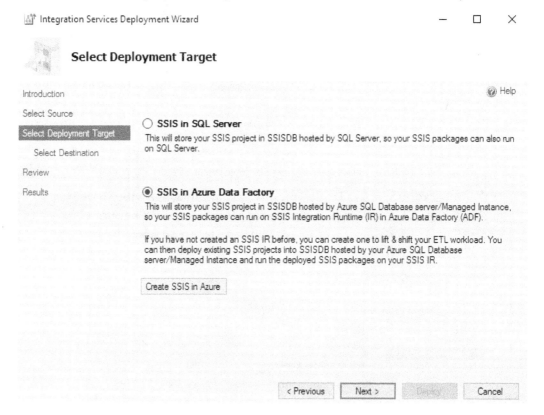

Figure 7.30 – Integration Services Deployment Wizard – Select Deployment Target

Click **Next** to go to the next step.

9. In the **Select Destination** tab, provide the Azure SQL Server admin user and password and click **Connect** to test the connection:

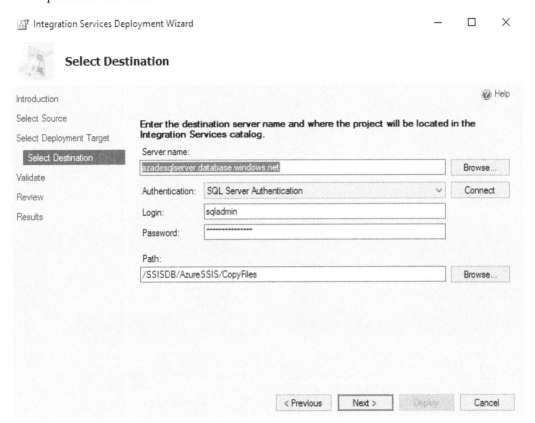

Figure 7.31 – Integration Services Deployment Wizard – Select Destination

10. After a successful connection, click **Next**. The deployment wizard will validate the package for issues, if any. After successful validation, click **Next** and then **Deploy** to complete the SSIS package deployment:

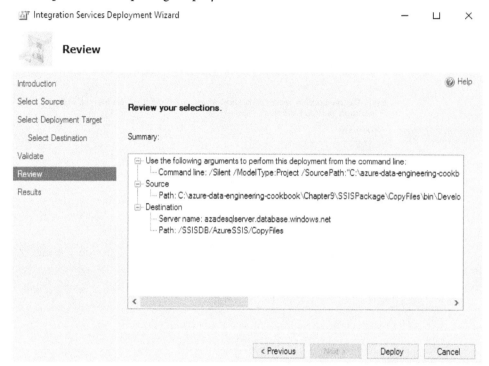

Figure 7.32 – Integration Services Deployment Wizard – Review and Deploy

After a successful deployment, close the wizard.

11. In the **Object Explorer** window in SQL Server Management Studio, refresh the AzureSSIS folder. Observe that the CopyFiles package is now listed under the AzureSSIS | Projects folder.

12. We'll now create an Azure Data Factory pipeline to execute the package we deployed to the SSISDB catalog. To do that, switch to the Azure portal. Open the Data Factory **Author** tab. Create a new pipeline called Pipeline-AzureSSIS-IR. Drag and drop the **Execute SSIS package** activity from the **Activities** to **General** section. In the **General** tab, name the activity Execute CopyFiles SSIS Package. Switch to the **Settings** tab. Select AzureSSISIR from the **Azure-SSISIR** drop-down list. Select the package location as SSISDB. Select AzureSSIS for **Folder**. If you don't see the folder name in the drop-down list, click **Refresh**. Select **CopyFiles** from the **Project** dropdown. Select CopyFiles.dtsx from the **Package** dropdown:

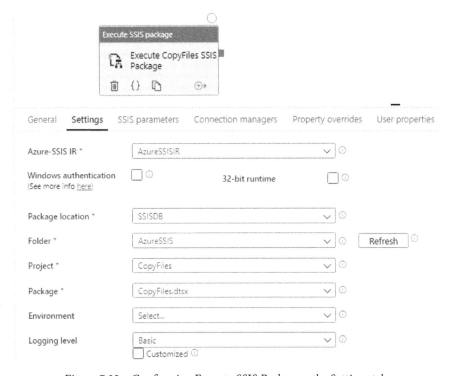

Figure 7.33 – Configuring Execute SSIS Package – the Settings tab

13. Switch to the **SSIS parameters** tab. Provide the **StorageAccountKey** and **StorageAccountName** values for the Azure storage account:

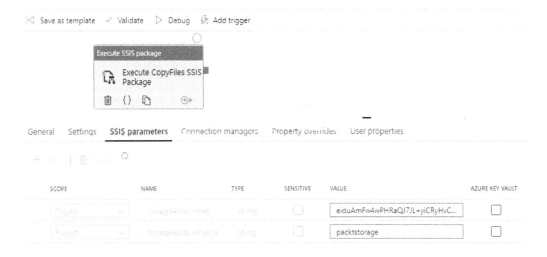

Figure 7.34 – Configuring Execute SSIS Package – SSIS parameters

> **Note**
>
> Make sure that the storage account you use has the `orders` container. You'll have to upload the `~/Chapter07/Data/orders1.txt` file to the `orders/datain` folder.

14. Click **Publish all** to save your work. Click **Debug** to run the package. Once the package is complete, you should get an output as shown in the following screenshot:

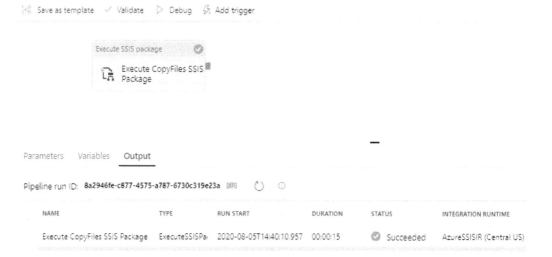

Figure 7.35 – Viewing the output

Observe that the IR used is `AzureSSISIR`.

> **Note**
>
> To stop the `Azure-SSIS` IR, navigate to **Data Factory | Manage | Integration runtimes**.
>
> Hover the mouse over the `Azure-SSIS` IR and click the **Pause** button.

Executing an SSIS package with an on-premises data store

We may have SSIS packages accessing an on-premises data source or the destination; for example, we may have files in an on-premises file store to be uploaded to Azure SQL Database, or we may have an on-premises database as the source. In such cases, we require the Azure SSIS IR to connect to the on-premises data store. There are two ways to do that:

- Configuring Azure SSIS to connect to on-premises using a Point-to-Site VPN, Site-to-Site VPN, or ExpressRoute. There are three steps to this:

  1) Set up a Point-to-Site VPN, Site-to-Site VPN, or ExpressRoute between on-premises and Azure.

  2) Create a virtual network and join the Azure SSIS IR with the virtual network.

  3) Create a virtual network gateway.

- Configure Azure SSIS to use a self-hosted IR as a proxy to connect to on-premises.

In this recipe, we'll explore the second option. To get more information on option 1, you can check out `https://docs.microsoft.com/en-us/azure/data-factory/join-azure-ssis-integration-runtime-virtual-network`.

Getting ready

To get started, do the following:

1. Log in to `https://portal.azure.com` using your Azure credentials.

2. Open a new PowerShell prompt. Execute the following command to log in to your Azure account from PowerShell:

   ```
   Connect-AzAccount
   ```

3. You will need an existing Data Factory account. If you don't have one, create one by executing the `~/azure-data-engineering-cookbook\Chapter04\3_CreatingAzureDataFactory.ps1` PowerShell script.

4. You will need a self-hosted IR. If you don't have an existing one, please follow the *Configuring a self-hosted IR* recipe to create a new one.

5. You will need an Azure SSIS IR. If you don't have an existing one, please follow the *Migrating an SSIS package to Azure Data Factory* recipe to create a new one.

6. You will need a storage account. You can create one by running the ~/Chapter06/2_UploadDataToAzureStorage.ps1 script.

7. You will need an Azure SQL database. You can create one by executing the ~/Chapter02/1_ProvisioningAzureSQLDB.ps1 PowerShell script.

How to do it...

In this recipe, we'll run an SSIS package to insert the data text files in an on-premises system into an Azure SQL database. The package we'll use is available at ~/Chapter07/ImportIntoAzureSQLDB.

Unlike the previous recipe where we deployed a package to SSISDB, in this recipe, we'll use PackageStore as the deployment method. PackageStore is an on-premises or Azure file share folder that contains the packages and required configuration files to be executed.

Let's start with creating a package store:

1. Execute the following PowerShell script to create a new storage file share:

```
# Get the storage account object
$str = Get-AzStorageAccount -ResourceGroupName packtade
-Name packtstorage
# Create the file share
New-AzStorageShare -Name ssisfileshare -Context $str.
Context
# Set the file share quota/size
Set-AzStorageShareQuota -ShareName ssisfileshare -Quota 1
-Context $str.Context
#upload SSIS package dtsx file to the file share
Set-AzStorageFileContent -ShareName ssisfileshare
-Source .\azure-data-engineering-cookbook\Chapter07\
ImportIntoAzureSQLDB\ImportIntoAzureSQLDB\
ImportIntoAzureSQLDB.dtsx -Path "/" -Context $str.Context
-Force

#upload SSIS package config file to the file share
Set-AzStorageFileContent -ShareName ssisfileshare
-Source .\azure-data-engineering-cookbook\Chapter07\
ImportIntoAzureSQLDB\ImportIntoAzureSQLDB\
ImportIntoAzureSQLDB.dtsconfig -Path "/" -Context $str.
Context -Force
```

The preceding script uses `Get-AzureStorageAccount` to get the storage object for an existing storage account. You'll need to change the `ResourceGroupName` and `Name` parameter values as per your environment.

The script uses the `New-AzStorageShare` cmdlet to create a new file share in the given storage account. The script sets the Azure file share size or quota to 1 GB. The `Set-AzStorageFileContent` command uploads the `dtsx` and `dtsconfig` files to the file share.

`ImportIntoAzureSQLDB.dtsx` is the SSIS package that creates the `orders` table if it doesn't exist, and then uploads the data from the `~/Chapter07/Data/Orders1.txt` file to the `orders` table in the Azure SQL database.

The package has the following parameters:

a) `azuresqlserver`: The default value is `azadesqlserver.database.windows.net`.

b) `azuresqldatabase`: The default value is `azadesqldb`.

c) `sqluser`: The default value is `sqladmin`.

d) `sqlpassword`: The default value is `Sql@Server@1234`.

e) `LocalPath`: The default value is `C:\azure-data-engineering-cookbook\Chapter07\Data\orders1.txt`.

Parameter values can be changed in the `ImportIntoAzureSQLDB.dtsconfig` file. For example, to change the name of the Azure SQL server, search for the default `azadesqlserver.database.windows.net` value and replace it accordingly.

2. We'll now create the Azure Storage linked service. To do that, open the `~/Chapter06/PacktStorageLinkedServiceDefinition.txt` file. Modify the account name and account key as per your storage account. Save and close the file. Execute the following command to create the linked service:

```
Set-AzDataFactoryV2LinkedService -Name
PacktStorageLinkedService -DefinitionFile .\
azure-data-engineering-cookbook\Chapter06\
PacktStorageLinkedServiceDefinition.txt -DataFactoryName
packtdatafactory -ResourceGroupName packtade
```

3. We'll now modify the Azure SSIS IR to use a self-hosted IR as a proxy to connect to the on-premises data store and configure the package store. To do that, open Data Factory and navigate to **Manage | Integration runtimes**. Hover the mouse over AzureSSISIR, and then select the pause button:

Figure 7.36 – Stopping the Azure SSIS IR

> **Note**
>
> We are modifying an existing Azure SSIS IR, and therefore it's required to stop the IR before making configuration changes. If you are creating a new IR, you can skip this step.

4. Click **Stop** in the **Confirmation** dialog box to stop the IR. When the IR is stopped, click on AzureSSISIR to open the **Edit integration runtime** dialog. Click **Continue** on the **General settings** tab to go to the next page. On the **Deployment settings** page, check the **Create package stores to manage your packages that are deployed into file system/Azure Files/SQL Server database (MSDB) hosted by Azure SQL Database Managed Instance** option:

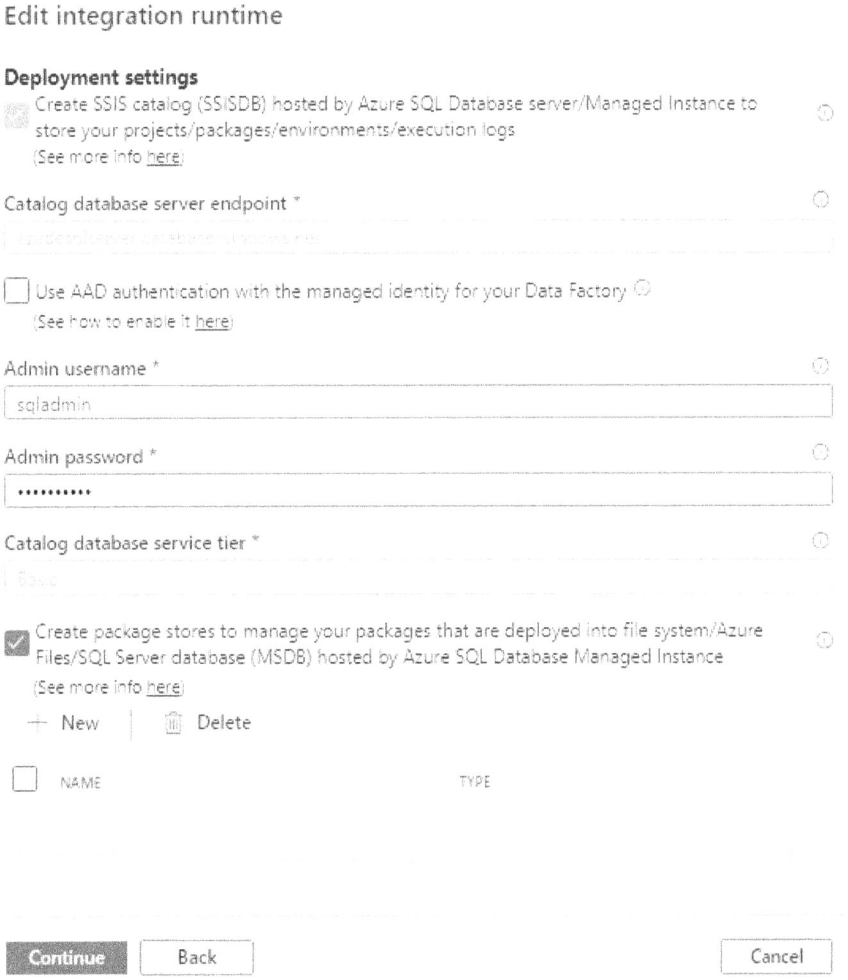

Edit integration runtime

Deployment settings

Create SSIS catalog (SSISDB) hosted by Azure SQL Database server/Managed Instance to store your projects/packages/environments/execution logs
(See more info here)

Catalog database server endpoint *

Use AAD authentication with the managed identity for your Data Factory
(See how to enable it here)

Admin username *

sqladmin

Admin password *

··········

Catalog database service tier *

Create package stores to manage your packages that are deployed into file system/Azure Files/SQL Server database (MSDB) hosted by Azure SQL Database Managed Instance
(See more info here)

+ New   🗑 Delete

NAME   TYPE

Continue   Back   Cancel

Figure 7.37 – Selecting package stores

5. Click **+ New** to add a new package store. On the **Add package store** page, set the package store name as `SSISStore`. In the **Package store linked service** dropdown, click **New**. On the **New linked service** page, set the name as `PackageStoreLinkedService` and for **File Share**, select `ssisfileshare`, which we created earlier:

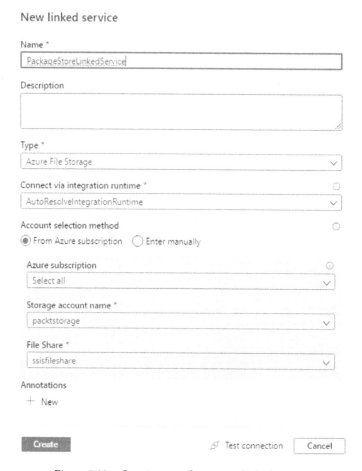

Figure 7.38 – Creating a package store linked service

6. Click **Create** to continue. The linked service will be created, and you'll be brought back to the **Add package store** page:

Figure 7.39 – Adding a package store

7.   Click **Add** to continue. You'll be brought back to the **Edit integration runtime |**
     Deployment settings page. Observe that SSISStore is now listed as the package
     store. Click **Continue** to go to the next step. On the **Advanced settings** page, check
     the **Set up Self-Hosted Integration Runtime as a proxy for your Azure-SSIS**
     Integration Runtime option. Select the self-hosted runtime and the storage linked
     service created earlier. Set the **Staging path** value as azuressisir:

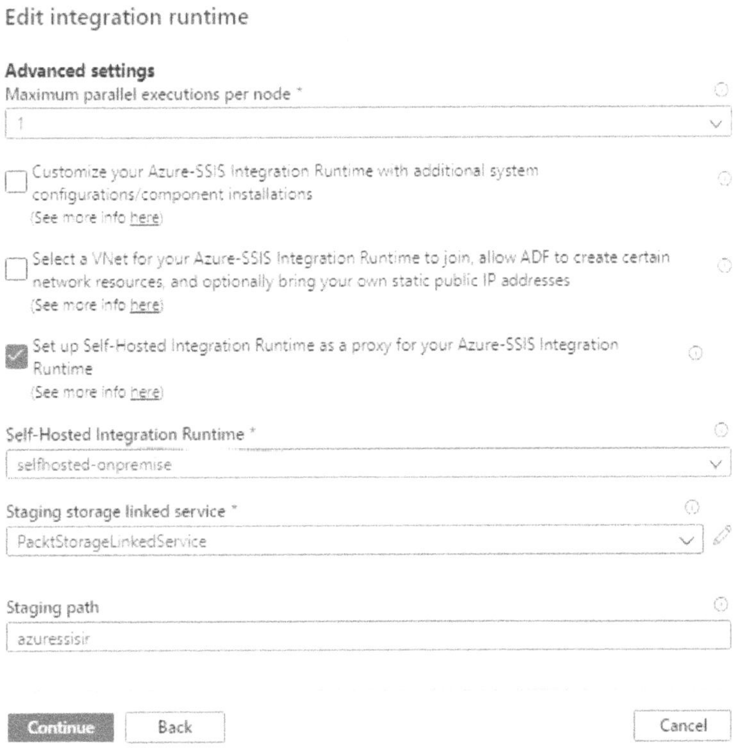

Edit integration runtime

Advanced settings

Maximum parallel executions per node *

1

☐ Customize your Azure-SSIS Integration Runtime with additional system
configurations/component installations
(See more info here)

☐ Select a VNet for your Azure-SSIS Integration Runtime to join, allow ADF to create certain
network resources, and optionally bring your own static public IP addresses
(See more info here)

☑ Set up Self-Hosted Integration Runtime as a proxy for your Azure-SSIS Integration
Runtime
(See more info here)

Self-Hosted Integration Runtime *

selfhosted-onpremise

Staging storage linked service *

PacktStorageLinkedService

Staging path

azuressisir

| Continue | Back |

Cancel

Figure 7.40 – Setting up a self-hosted IR as a proxy

The proxy works by first fetching the data from the on-premises store to the staging path using a self-hosted IR, and then moving the data from the Azure Storage staging area to the destination using Azure SSIS IR.

8. Click **Continue**, and then click **Update** to save the changes. Start the Azure SSIS IR.

9. Navigate to the **Data Factory Author** page and create a new pipeline called `Pipeline-SelfHosted-Proxy`. Drag and drop the **Execute SSIS Package** activity from the **Activities | General** section. In the **General** tab, name the activity `Import Data from On-Premise`. In the **Settings** tab, select the `Azure-SSIS` IR. Select **Package store** for **Package location**. Select `SSISStore` for **Package store name**. Provide `ImportIntoAzureSQLDB` for **Package path**. The package path should be the package name without the `dtsx` extension. You can also select the package by clicking the **Browse file storage** button. Set `\\packtstorage.file.core.windows.net\ssisfileshare\ImportIntoAzureSQLDB.dtsConfig` as the **Configuration path** value. This is the path to the `dtsconfig` file. You can also use the **Browse file storage** button to select the configuration file. Leave the **Encryption password** field blank. The encryption password is the one that is provided in the package properties if the package **Protection Level** value is set to `EncryptSensitiveWithPassword` or `EncryptAllWithPassword`. The SSIS package here has **Protection Level** as `DontSaveSensitive` and therefore, no password is required. We can change **Logging level** as required and save logs to a custom path. However, the values are left as default for this recipe:

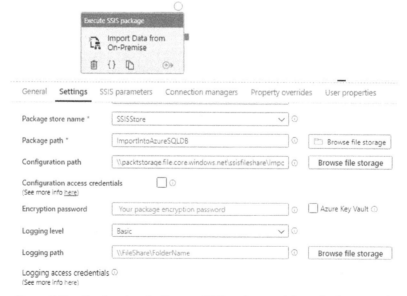

Figure 7.41 – Configuring the Execute SSIS package activity – the Settings tab

Note that when using a self-hosted IR as a proxy for Azure SSIS to connect to an on-premises data store, we need to set the **Connection Manager** (flat file/OLEDB/ODBC) **ConnectyByProxy** property to **True** in the SSIS package (as shown in the following screenshot):

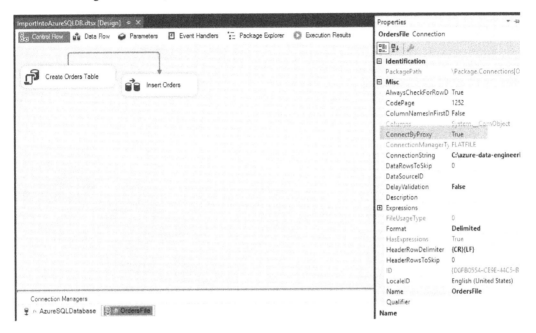

Figure 7.42 – ImportIntoAzureSQLDB SSIS package ConnectByProxy setting

10. The pipeline configuration is complete. Click **Publish all** to save your work, and then click **Debug** to run the pipeline. When the pipeline completes, you'll get an output as shown in the following screenshot:

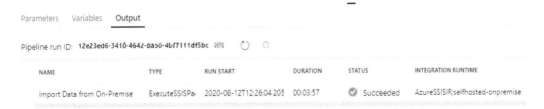

Figure 7.43 – Viewing the output

Observe that the task was performed by both the self-hosted and Azure SSIS IRs.

8
Deploying Azure Data Factory Pipelines

In previous chapters, we have talked about creating Azure Data Factory pipelines for different data engineering use cases, and also learned the different activities and data flows available in Azure Data Factory. In this chapter, we'll talk about continuous integration and deployment of Azure Data Factory pipelines.

We'll first learn to manually deploy pipelines across different environments, such as development, testing, and production, and will then use Azure DevOps to automate the deployment process.

In this chapter, we'll cover the following recipes:

- Configuring the development, test, and production environments
- Deploying Azure Data Factory pipelines using the Azure portal and ARM templates
- Automating the Azure Data Factory pipeline deployment using Azure DevOps

Technical requirements

For this chapter, the following is required:

- A Microsoft Azure subscription
- PowerShell 7
- Microsoft Azure PowerShell

Configuring the development, test, and production environments

A typical software development environment consists of three environments, **development**, **test** (or staging), and **production**. A development environment is where developers create and unit-test the code; in our case, Azure Data Factory pipelines.

When the development is complete, the pipeline is deployed (moved) to the test environment. In the test environment, testing teams perform tests to validate the pipeline against the business requirements. Any bugs raised are noted and passed on to the developers. The developers then fix the bugs in the development environment and push the changes to the test environment. This goes on until the pipeline passes all of the validation and testing. The pipeline is then pushed to the production environment where it works on the actual data.

In some cases, there can be additional environments such as performance testing, which can be used to performance-test the pipeline before it is deployed to production.

In this recipe, we'll configure the development, test, and production environments.

We'll create a pipeline to copy data from an Azure storage account to an Azure SQL database using the copy activity in the development environment, and integrate it with the source control.

Getting ready

To get started, open a PowerShell console window, run the `Connect-Azaccount` command, and follow the instructions to log in to your Azure account.

How to do it...

Let's start by creating three resource groups, one for each (development, test, and production) environment:

1. Execute the following PowerShell commands to create the three resource groups, one for each environment (development, test, and production):

    ```
    # Development Resource group - packt-rg-dev
    New-AzResourceGroup -Name packt-rg-dev -Location
    centralus
    # Test Resource group - packt-rg-dev
    New-AzResourceGroup -Name packt-rg-tst -Location
    centralus
    # Production Resource group - packt-rg-prd
    New-AzResourceGroup -Name packt-rg-prd -Location
    centralus
    ```

    The preceding script creates three resource groups: `packt-rg-dev` for the development environment, `packt-rg-tst` for the test environment, and `packt-rg-prd` for the production environment.

2. Execute the following PowerShell script to create three Data Factory services, one for each environment:

    ```
    # Create the development data factory account in the
    development resource group
    New-AzDataFactoryV2 -ResourceGroupName packt-rg-dev -Name
    packt-df-dev -Location centralus
    # Create the test data factory account in the test
    resource group
    New-AzDataFactoryV2 -ResourceGroupName packt-rg-tst -Name
    packt-df-tst -Location centralus
    # Create the production data factory account in the
    production resource group
    New-AzDataFactoryV2 -ResourceGroupName packt-rg-prd -Name
    packt-df-prd -Location centralus
    ```

    The preceding code creates three Data Factory accounts, `packt-df-dev`, `packt-df-tst`, and `packt-df-prd`, in their respective resource groups.

3. Execute the following PowerShell scripts to create three Azure SQL databases, one for each environment:

```
#Create the development Azure SQL Server and the
database.
C:\azure-data-engineering-cookbook\Chapter06\1_
ProvisioningAzureSQLDB.ps1 -resourcegroup packt-rg-dev
-servername packt-sqlserver-dev -databasename azsqldb
-password Sql@Server@123 -location centralus
```

```
#Create the test Azure SQL Server and the database.
C:\azure-data-engineering-cookbook\Chapter06\1_
ProvisioningAzureSQLDB.ps1 -resourcegroup packt-rg-tst
-servername packt-sqlserver-tst -databasename azsqldb
-password Sql@Server@123 -location centralus
```

```
#Create the production Azure SQL Server and the database.
C:\azure-data-engineering-cookbook\Chapter06\1_
ProvisioningAzureSQLDB.ps1 -resourcegroup packt-rg-prd
-servername packt-sqlserver-prd -databasename azsqldb
-password Sql@Server@123 -location centralus
```

The preceding command creates three Azure SQL Server instances, `packt-sqlserver-dev`, `packt-sqlserver-tst`, and `packt-sqlserver-prd`, each with one database, `azsqldb`. The Azure SQL databases are created in the **Basic Service Tier**.

4. Execute the following PowerShell commands to create an Azure storage account for each environment, and upload the sample file to the storage accounts:

```
#Create an Azure storage account for the development
environment and upload sample file.
 C:\azure-data-engineering-cookbook\Chapter06\2_
UploadDatatoAzureStorage.ps1 -resourcegroupname packt-
rg-dev -storageaccountname packtstoragedev -location
centralus -datadirectory C:\azure-data-engineering-
cookbook\Chapter06\Data\ -createstorageaccount $true
-uploadfiles $true
```
```
#Create an Azure storage account for the test environment
and upload sample file.
 C:\azure-data-engineering-cookbook\Chapter06\2_
UploadDatatoAzureStorage.ps1 -resourcegroupname packt-
rg-tst -storageaccountname packtstoragetst -location
```

```
centralus -datadirectory C:\azure-data-engineering-
cookbook\Chapter06\Data\ -createstorageaccount $true
-uploadfiles $true

#Create an Azure storage account for the production
environment and upload sample file.
 C:\azure-data-engineering-cookbook\Chapter06\2_
UploadDatatoAzureStorage.ps1 -resourcegroupname packt-
rg-prd -storageaccountname packtstorageprd -location
centralus -datadirectory C:\azure-data-engineering-
cookbook\Chapter06\Data\ -createstorageaccount $true
-uploadfiles $true
```

The preceding commands create three Azure storage accounts,
`packtstoragedev`, `packtstoragetst`, and `packtstorageprd`.

5. We'll use Azure Key Vault to store the Azure SQL database and Azure storage
   account credentials. We'll then use Azure Key Vault when connecting to these
   services from Data Factory-linked services. Execute the following command to
   create three Azure Key Vault accounts, one for each storage account, and then create
   a secret key to store the credentials for the Azure SQL databases and Azure storage
   accounts. The following PowerShell command will create an Azure Key Vault
   instance for the development environment:

```
#Create Azure Key Vault for the development environment.
New-AzKeyVault -Name packt-keyvault-dev
-ResourceGroupName packt-rg-dev -Location centralus
```

6. Execute the following command to create a secret in Azure Key Vault to store the
   Azure storage account connection string:

```
#Get Azure storage account connection string
$sacs = (Get-AzStorageAccount -ResourceGroupName packt-
rg-dev -Name packtstoragedev).Context.ConnectionString
#Create secret in Azure Key Vault to store the key.
Set-AzKeyVaultSecret -VaultName packt-keyvault-dev -Name
azure-storage-connectionstring -SecretValue (ConvertTo-
SecureString $sacs  -AsPlainText -Force)
```

7. Execute the following command to save the Azure SQL Server name as a secret key in Azure Key Vault:

```
$azuresqldbconnectionstring =
"Server=tcp:packt-sqlserver-dev.database.windows.
net,1433;
Initial Catalog=azsqldb;Persist Security Info=False;User
ID=sqladmin;Password=Sql@Server@123;
MultipleActiveResultSets=False;Encrypt=True;
TrustServerCertificate=False;Connection Timeout=30;"
```

```
Set-AzKeyVaultSecret -VaultName packt-keyvault-dev
-Name azure-sqldb-connectionstring -SecretValue
(ConvertTo-SecureString $azuresqldbconnectionstring
-AsPlainText -Force)
```

The preceding command assigns the connections string for the packt-sqlserver-dev Azure SQL database to the $azuresqldbconnectionstring variable, and then stores the connection string as a secret key, azure-sqldb-connectionstring, in Azure Key Vault.

8. Execute the following command to store the environment name in Azure Key Vault. We'll use the environment later in the chapter when automating the Data Factory pipeline deployment using Azure DevOps:

```
Set-AzKeyVaultSecret -VaultName packt-keyvault-dev -Name
environment-name -SecretValue (ConvertTo-SecureString
"dev" -AsPlainText -Force)
```

The preceding command will create an environment-name secret key in Azure Key Vault with an environment value of dev. The values for the test and production environments will be tst and prd respectively.

9. Execute the following command to give access to the development data factory to read the secrets from the development Azure Key Vault:

```
#Get the data factory PrincipalId
```

```
$dfPrincipalId = (Get-AzDataFactoryV2 -ResourceGroupName
packt-rg-dev -Name packt-df-dev).Identity.PrincipalId
```

```
#Grant Get and list access on Azure Key Vault secret to
data factory
```

```
Set-AzKeyVaultAccessPolicy -VaultName packt-keyvault-dev
-ResourceGroupName packt-rg-dev -ObjectId $dfPrincipalId
-PermissionsToSecrets Get,List
```

The preceding command gives access to Azure Data Factory to read and list the secret keys from the `packt-keyvault-dev` Azure Key Vault. Replace the resource group name, Azure Key Vault name, Data Factory name, and Azure SQL database name with those of the testing and production environments, and run the preceding commands to configure Azure Key Vault for the testing and production environments. Keep the Azure Key Vault secret name the same for all environments. This will help when creating a linked service as we can select different Azure Key Vault instances depending on the environment; however, by using the same secret key name, we will not need to change it. After completing the preceding steps, you should have three environments with their own sets of services defined.

The following is the development environment:

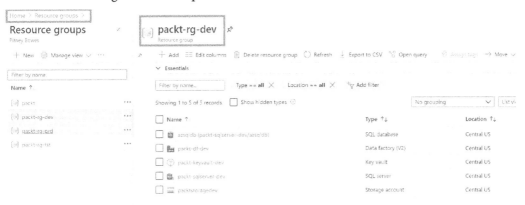

Figure 8.1 – Development environment – Resource group – packt-rg-dev

The following is the test environment:

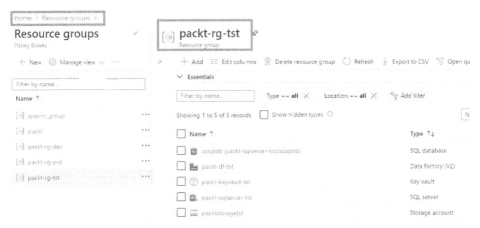

Figure 8.2 – Test environment – Resource group – packt-rg-tst

The following is the production environment:

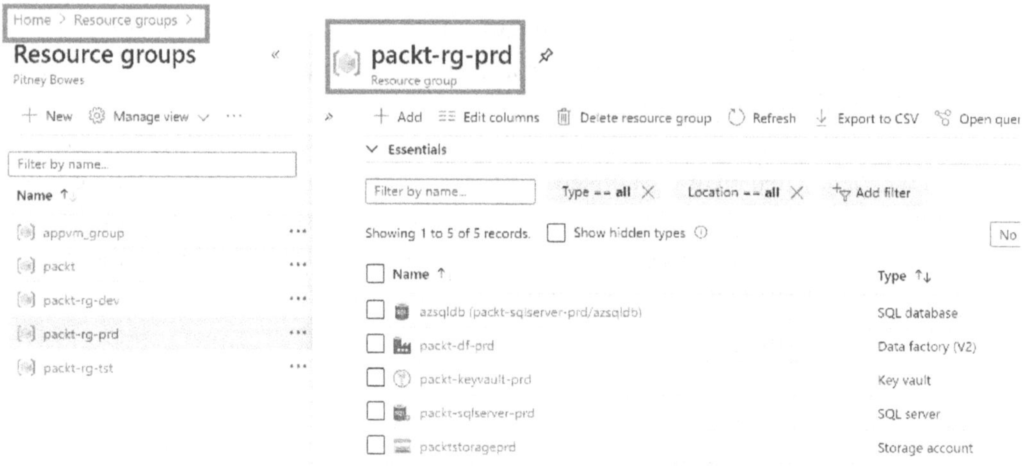

Figure 8.3 – Production environment – Resource group – packt-rg-prd

Let's now create a pipeline to insert data from a file in Azure Blob storage in Azure SQL Database in the development environment.

10. Execute the following PowerShell command to create the Azure Key Vault linked service:

```
Set-AzDataFactoryV2LinkedService -Name
KeyVaultLinkedService -DefinitionFile C:\azure-data-
engineering-cookbook\Chapter08\KeyVaultLSDefinitionFile.
txt -ResourceGroupName packt-rg-dev -DataFactoryName
packt-df-dev
```

The preceding command uses the definition file to create KeyVaultLinkedService in the packt-df-dev data factory.

11. Execute the following PowerShell command to create the Azure Blob storage linked service:

```
Set-AzDataFactoryV2LinkedService -Name
StorageLinkedService -DefinitionFile C:\azure-data-
engineering-cookbook\Chapter08\StorageLSDefinitionFile.
txt -ResourceGroupName packt-rg-dev -DataFactoryName
packt-df-dev
```

The preceding command uses the definition file to create StorageLinkedService in the packt-df-dev data factory.

12. Execute the following command to create an Azure SQL database linked service:

```
Set-AzDataFactoryV2LinkedService -Name
SQLDatabaseLinkedService -DefinitionFile C:\
azure-data-engineering-cookbook\Chapter08\
SQLDatabaseLSDefinitionFile.txt -ResourceGroupName packt-
rg-dev -DataFactoryName packt-df-dev
```

The preceding command creates `SQLDatabaseLinkedService` using the definition file in the `packt-df-dev` data factory.

13. Let's create two datasets, one from the file in Azure Blob storage and another one from the table in the Azure SQL database. Execute the following PowerShell command to create the `Orders` dataset to read the `Orders1.txt` file from Azure Blob storage using a definition file:

```
Set-AzDataFactoryV2Dataset -Name Orders -DefinitionFile
C:\azure-data-engineering-cookbook\Chapter08\
OrdersDSDefinitionFile.txt -ResourceGroupName packt-rg-
dev -DataFactoryName packt-df-dev
```

The preceding command creates the `Orders` dataset in the `packt-df-dev` data factory using a definition file. If you used a different linked service name in *step 7*, then replace the linked service name in the `OrdersDSDefinitionFile.txt` file under the `LinkedServiceName` | `referenceName` element.

14. Execute the following PowerShell command to create the `OrdersTable` dataset for the `Orders` table in the Azure SQL database:

```
Set-AzDataFactoryV2Dataset -Name OrdersTable
-DefinitionFile C:\azure-data-engineering-
cookbook\Chapter08\OrdersTableDSDefinitionFile.txt
-ResourceGroupName packt-rg-dev -DataFactoryName packt-
df-dev
```

The preceding command creates the `OrdersTable` dataset in the `packt-df-dev` data factory using a definition file. If you used a different linked service name in *step 8*, then replace the linked service name in the `OrdersTableDSDefinitionFile.txt` file under the `LinkedServiceName` | `referenceName` element.

15. With the linked services and datasets in place, let's create the pipeline. Execute the following PowerShell command to create a pipeline to copy data from the Order1. txt file in Azure Blob storage to the Orders table in the Azure SQL database:

```
Set-AzDataFactoryV2Pipeline -Name Pipeline-Copy-Data
-DefinitionFile C:\azure-data-engineering-cookbook\
Chapter08\PipelineDefinitionFile.txt -ResourceGroupName
packt-rg-dev -DataFactoryName packt-df-dev
```

The preceding command creates a Pipeline-Copy-Data pipeline using the definition file in the packt-df-dev data factory.

16. This completes our setup. We created three environments, development, test, and production, and the required services for those three environments. We then created a pipeline in the development environment. Let's head over to the packt-df-dev data factory author page in the Azure portal and run the pipeline:

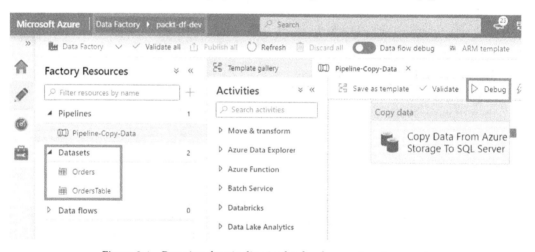

Figure 8.4 – Running the pipeline in the development environment

17. Click **Debug** to run the pipeline. You should get an output as shown in the following screenshot when the pipeline completes:

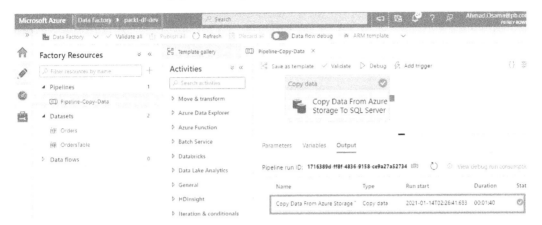

Figure 8.5 – Running the pipeline in the development environment – status

The pipeline reads the data from the `orders1.txt` file in the `packtstoragedev` Azure storage account and loads the data into the `Orders` table in the `packt-sqlserver-dev` Azure SQL database. The table is automatically created if it doesn't already exist.

In the next recipe, we'll learn to integrate the pipeline with source control and manually deploy the pipeline to the test and production environments.

Deploying Azure Data Factory pipelines using the Azure portal and ARM templates

In this recipe, we'll integrate the pipeline we created in the previous recipe with source control, and we'll learn how to deploy the pipeline to the test and production environment manually using the Azure portal and ARM templates.

Getting ready

To get started, you will need to do the following:

1. Open the Azure portal at `https://portal.azure.com` and log in to your Azure account.

2. Create an Azure DevOps organization and project. We'll use this project to source control the **Data Factory** pipeline. Refer to the following link on how to create an organization and a project in Azure DevOps: `https://docs.microsoft.com/en-us/azure/devops/organizations/accounts/create-organization?view=azure-devops`.

How to do it...

Let's start integrating the data factory with source control. There are two source control options available with Azure Data Factory – **Azure DevOps Git** and **GitHub**. We'll use the Azure DevOps Git as source control:

1. In the Azure portal, open the `packt-df-dev` author page and then open **Pipeline-Copy-Data**. From the **Data Factory** dropdown, select **Set up code repository**:

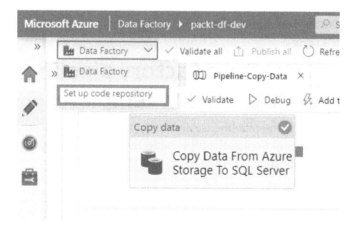

Figure 8.6 – Setting up a code repository

2. On the **Configure a repository** page, select the repository type as **Azure DevOps Git** and then select **Azure Active Directory** (if there is more than one Active Directory in your Azure account). In the **Azure DevOps Account** dropdown, select the Azure DevOps organization. The **Project name** dropdown will list out the available projects in the selected Azure DevOps organization. Select the one in which you want to keep the data factory code. In the **Repository name** section, check **Use existing** and then select the one available in the dropdown. If this is the new project, the existing repository name will be same as the project name. The collaboration branch is the one from where we'll publish the pipeline. Leave it as default (**master** branch) for now. Leave the rest of the options as the default values:

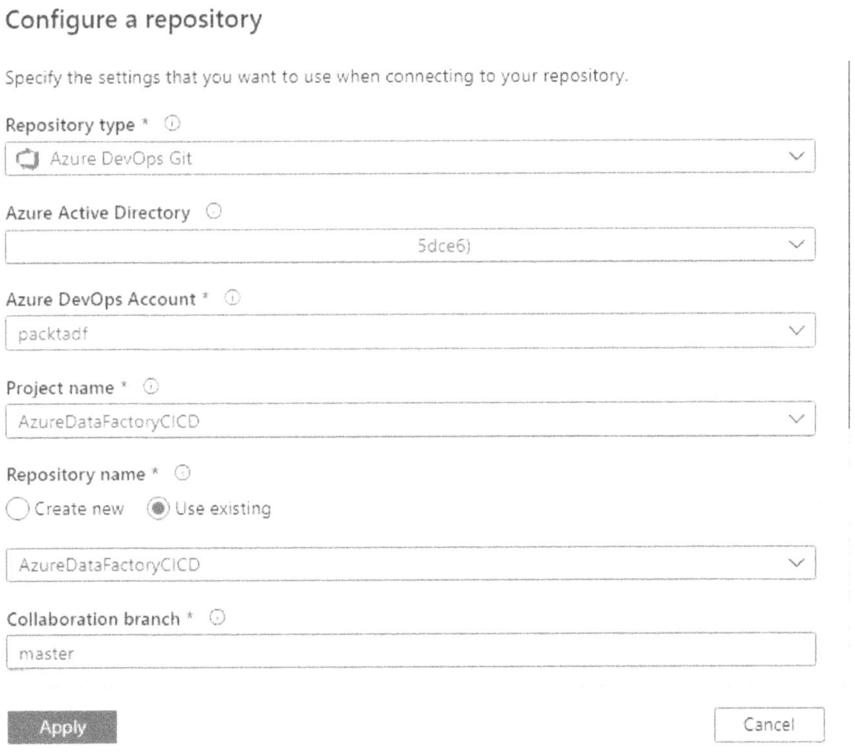

Configure a repository

Specify the settings that you want to use when connecting to your repository.

Repository type * ⓘ

⟳ Azure DevOps Git	⌄

Azure Active Directory ○

5dce6)	⌄

Azure DevOps Account * ⓘ

packtadf	⌄

Project name * ⓘ

AzureDataFactoryCICD	⌄

Repository name * ⓘ

○ Create new   ⦿ Use existing

AzureDataFactoryCICD	⌄

Collaboration branch * ⓘ

master

[Apply]   [Cancel]

Figure 8.7 – Setting up a code repository – Configuring Azure DevOps Git

As shown in the preceding screenshot, the code for the pipeline will be checked into the `packtdf` Azure DevOps account in the `AzureDataFactoryCICD` project.

3. Click **Apply** to save the configuration. The **Data Factory** pipeline will be connected to the source control and there'll be a popup in which you can select the working branch:

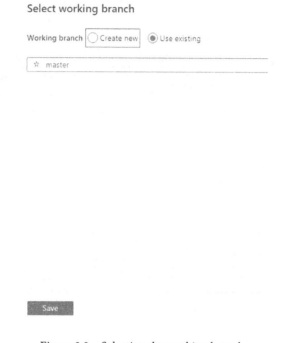

Figure 8.8 – Selecting the working branch

4. Select **master** (default) as the working branch and click **Save**.

5. Let's head over to Azure DevOps and review the content of the **master** branch:

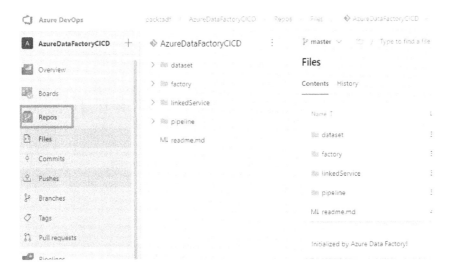

Figure 8.9 – Reviewing Azure DevOps content – Repos

In Azure DevOps, navigate to your project and select **Repos** from the left menu as shown in the preceding screenshot. Observe that the contents of the pipeline are converted to JSON code and committed to the **master** branch. There's a folder for each object in the data factory, dataset, factory, linked service, and pipeline. You can review the JSON files by navigating to the specific folder and clicking the JSON file. These JSON files are known as **Azure Resource Manager templates** (**ARM templates**) and are used to deploy infrastructure as code. In the next steps, we'll use ARM templates to deploy the pipeline to the test and production environments.

6. To deploy the pipeline to the test environment, we'll export the pipeline ARM template from the development environment and then import the template in the test environment. Navigate to the **Pipeline-Data-Copy** pipeline in the development data factory (**packt-df-dev**). Select **ARM template** from the top menu and then select **Export ARM template**:

Figure 8.10 – Exporting the ARM template

The process downloads the `arm_template.zip` file to your default `Download` folder. Navigate to the file location and unzip the `arm_template.zip` folder. The `arm_template.json` file contains the information required to create the pipeline and the required objects in JSON format. We'll now import this file into the test environment to deploy the pipeline in the test environment.

7. In the Azure portal, navigate to the test data factory (**packt-df-tst**) author section. Select the **ARM template** dropdown and then select **Import ARM template**. A new **Custom deployment** tab will open:

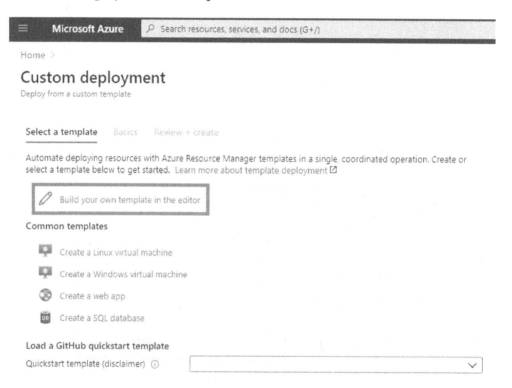

Figure 8.11 – Importing ARM template – Custom deployment

8. Select **Build your own template in the editor** option. On the **Edit template** page, select **Load file**. Navigate to `arm_template` (the unzipped folder from *step 6*) and select the `arm_template.json` file:

Figure 8.12 – Importing ARM template – Load arm_template.json file

The `arm_template` folder is loaded and displayed as shown in the preceding screenshot. Click **Save** to continue.

9. On the **Custom deployment** page, select **packt-rg-tst** for **Resource group** and
    packt-df-tst for **Factory Name**. The SQL database and Azure Storage connection
    strings are stored in Azure Key Vault. Modify the key vault URL to `https://`
    `packt-keyvault-tst.vault.azure.net/` to point to the test Azure Key
    Vault instance. We don't need to change the secret key name as we created the same
    secret key name across all the environments:

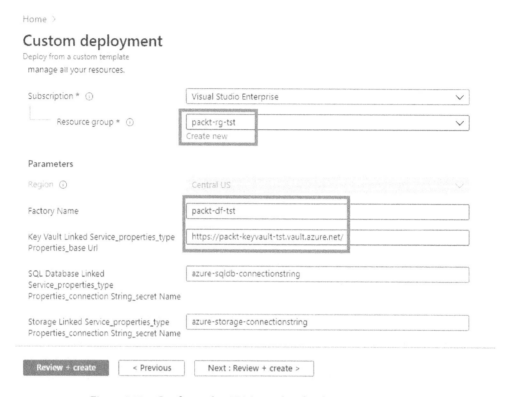

Figure 8.13 – Configure the ARM template for the test environment

10. Click **Review + create** and then click **Create** to import the ARM template. You should get the following screenshot, detailing the successful deployment of the pipeline:

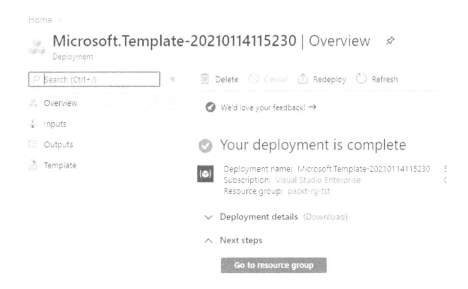

Figure 8.14 – Importing ARM template – Deployment successful

11. Navigate to the **packt-df-tst** data factory author section. Observe that the **Pipeline-Copy-Data** pipeline is successfully deployed in our test environment:

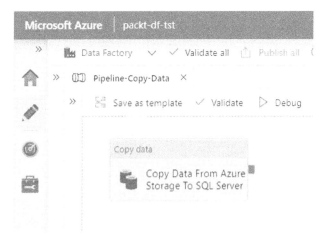

Figure 8.15 – Reviewing the pipeline in the test environment

12. Click **Debug** to run and test the pipeline.

We can deploy the pipeline to the production data factory `packt-df-prd` in similar ways by following *step 3*. We do need to change the parameters to point to the production data factory and Azure Key Vault.

In this recipe, we learned to integrate pipelines with Azure DevOps Git and to manually deploy the pipeline to the test and production environments. In the next recipe, we'll set up automatic pipeline deployment using a pipeline in Azure DevOps.

Automating Azure Data Factory pipeline deployment using Azure DevOps

In this recipe, we'll create an Azure DevOps pipeline to automatically deploy the data factory pipeline to the test and production environments. The Azure DevOps pipeline consists of a series of tasks to automate the manual deployment steps we performed in the previous recipe. To find out more about Azure DevOps pipelines, refer to `https://docs.microsoft.com/en-us/azure/devops/?view=azure-devops`.

Getting ready

To get started, you will need to do the following:

1. Open the Azure portal at `https://portal.azure.com` and log in to your Azure account.

2. Log in to `https://devops.azure.com`.

How to do it...

Let's get started by creating variable groups in Azure DevOps. Variable groups define the variable name and values for the test and production environment. We'll use the variable group to pass the variables values to pipeline task in later steps:

1. In the Azure DevOps portal, open the data factory project. From the left menu, select **Pipelines | Library**:

Figure 8.16 – Creating variable groups – Pipelines | Library

2. On the **Library** page, click on **+ Variable group**. On the **Variable group |**
    Properties page, provide `test` as the variable group name and turn on the
    Link secrets from an Azure key vault as variables option. Select your Azure
    subscription and then select the **packt-keyvault-tst** key vault. In the **Variables**
    section, click **Add** to open the **Choose secrets** pop-up window. In the pop-up
    window, select `environment-name`. You'll have to click the **Authorize** button to
    first grant access to Azure DevOps to get the subscription detail and then to allow
    Azure DevOps to read the Azure Key Vault secrets:

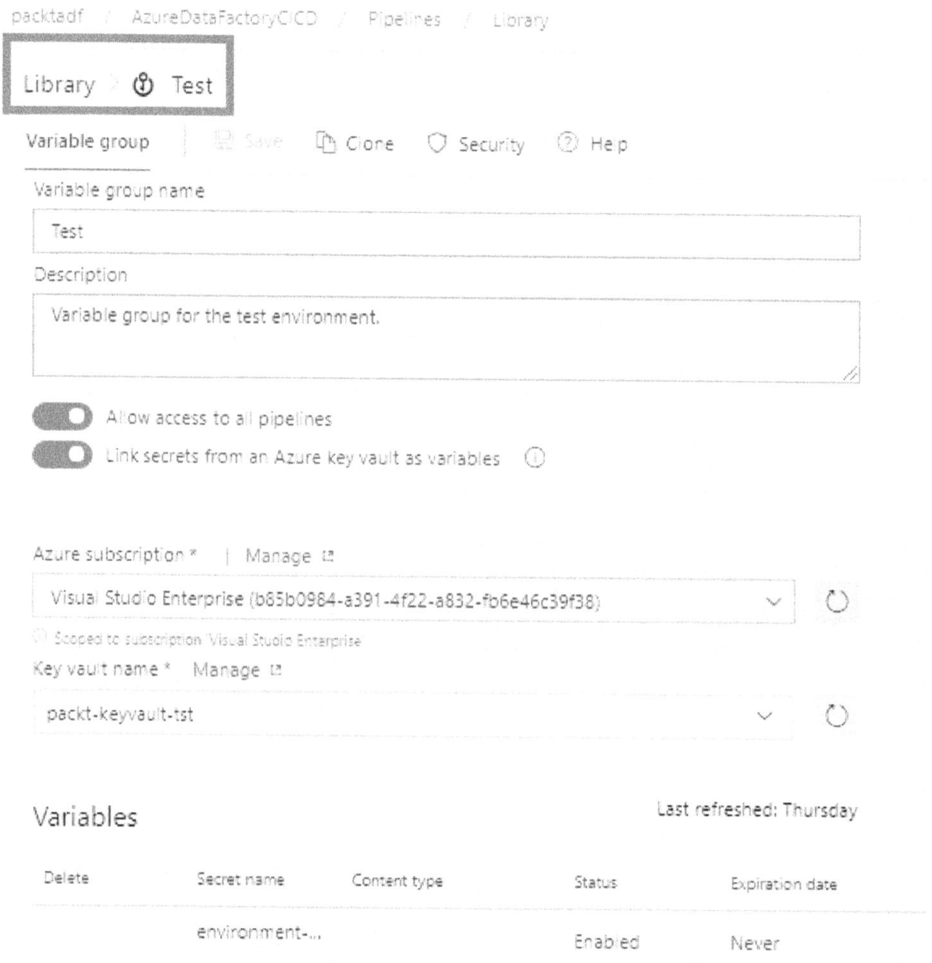

Figure 8.17 – Creating a variable group and adding variables for the test environment

3. Click **Save** to save the variable group configuration. Similarly, create a new variable group called `Production` to store the production variable values:

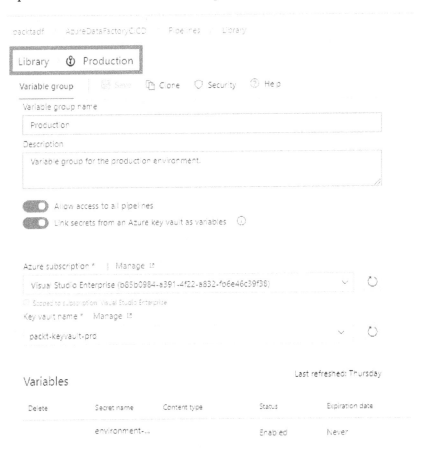

Figure 8.18 – Creating a variable group and adding variables for the production environment

Observe that in the **Production** variable group, the key vault selected is packt-keyvault-prd. The next step is to create the release pipeline.

4. In the Azure DevOps portal, open the data factory project. Select the **Pipelines** option from the left menu and select **Releases**:

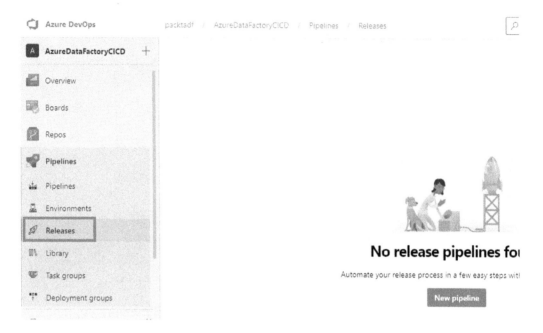

Figure 8.19 – Creating a new release pipeline

5. Select **New pipeline** to create a new release pipeline. On the **Select a template** page, select **Empty job**:

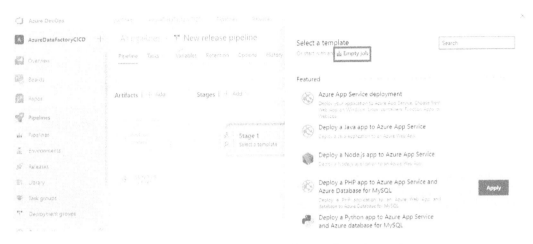

Figure 8.20 – Creating release pipeline – Select job template

6. A pipeline consists of multiple stages. Each stage consists of tasks to be performed to release the code and represents an environment. In the **Stage** window, provide Test as the stage name:

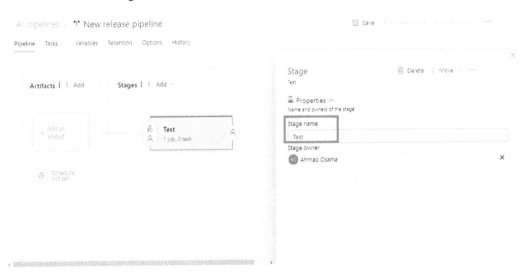

Figure 8.21 – Creating release pipeline – Creating new stage

Close the **Stage** window by clicking on the top-right close (**X**) button.

7. We'll add an artifact to the pipeline. Consider artifacts as the file or package (such as NuGet, npm, or Python) to be deployed to a given environment. In our case, the artifact is the ARM template stored in Azure DevOps Git in the `adf_publish` branch. The `adf_publish` branch contains the data factory pipeline ARM templates when we publish the changes. We'll look at this later in the recipe. In the **Release Pipeline Artifacts** section, click **+ Add**. In the **Add an artifact** window, select **Azure Repos** as **Source type**. Select the project and the repository. We created the project and repository in *step 1* of the second recipe in this chapter. Select **adf_publish** for the **Default branch** field:

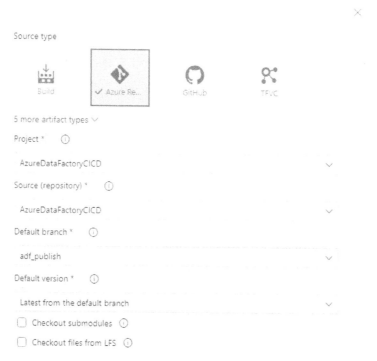

Figure 8.22 – Adding an artifact to the release pipeline

Leave the rest of the settings as their defaults. Scroll to the end of the page and click **Add**.

8. We'll add the task to the stages to deploy the selected artifact. In the **Stages** section, select **one job, zero task** to add the task to the **Test** stage. Alternatively, you can also click on the **Tasks** menu option and select the **Test** stage. In the **Tasks** section, click the + icon next to the **Agent job** section to add a new task:

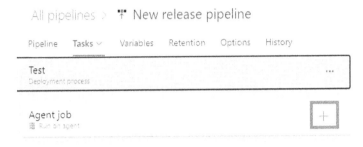

Figure 8.23 – Adding a new task to the Test stage

9. In the **Add tasks** page's search box, type ARM and then select the **ARM template deployment** task:

Figure 8.24 – Selecting the ARM template deployment task

The **ARM template deployment** task, as the name suggests, deploys an ARM template in the specified resource group.

10. Click **Add** to add the task to the **Test** stage. Click on the **ARM template deployment** task under the agent job to configure the parameters. Select the latest generally available task version. Modify the display name to ARM Template deployment: Copy-Data Pipeline. For the deployment scope, select **Resource Group**. Select the available Azure resource manager connection. If you don't have one, select the subscription in which you want to deploy the pipeline and then click the **Authorize** button. In the **Subscription** dropdown, select your Azure subscription:

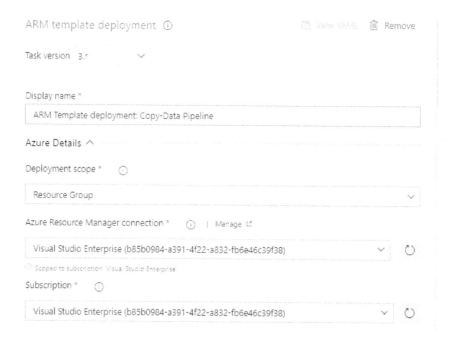

Figure 8.25 – Configuring ARM template deployment task – part 1

11. Scroll down further to configure the rest of the settings. For **Action**, select **Create or update resource group**. Choose **packt-rg-$(environment-name)** for **Resource group**. This makes the resource group name generic, as the environment name is picked up from the environment-name parameter from the variable group. Later in the recipe, we'll link the variable groups to the corresponding stages. Select **Central US** for the location, as all of our resource groups are in the **Central US** region:

Figure 8.26 – Configuring ARM template deployment task – part 2

12. Scroll down to the **Template** section to define the ARM template JSON file to be deployed. Select **Linked artifact** for **Template location**. To select the ARM template, click the ellipsis button next to the **Template** text box to browse the available ARM templates in the linked artifact. On the **Select a file or folder** pop-up page, expand the `AzureDataFactoryCICD` repository (the repository name may be different in your case), then expand the `packt-df-dev` folder, and select the `ARMTemplateForFactory.json` file:

Figure 8.27 – Selecting the data factory pipeline ARM template

Click **OK** to select the file.

13. To select the template parameters, select the ellipsis button next to the **Template parameters** text box and select the `ARMTemplateParametersForFactory.json` file.

14. In the **Override template parameters** section, we'll modify the parameters to make them generic. To do that, click on the ellipsis button to open the **Override template parameters** pop-up page. Modify the **factoryName** parameter to `packt-df-$(environment-name)` and **KeyVaultLinkedService URL** to `https://packt-keyvault-$(environment-name).vault.azure.net/:`

Figure 8.28 – Overriding template parameters

Click **OK** to continue.

The template configuration should be as shown:

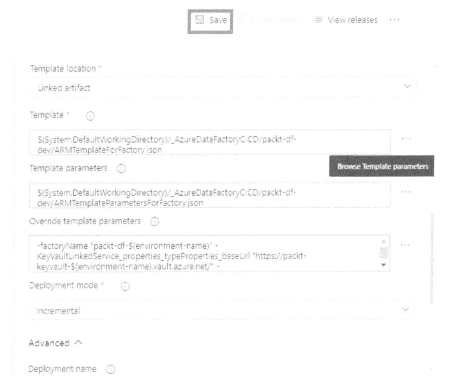

Figure 8.29 – Configuring ARM template deployment task – part 3

Leave the rest of the settings as their defaults. Click **Save** to save the configuration. This completes the configuration for the **Test** stage.

Note

If a pipeline has triggers defined, then the triggers are to be stopped before deployment and started after the deployment, otherwise the deployment will fail. To do this, add a pre- and post-deployment (PowerShell) task to run a script to stop and start the triggers. The script and the steps are available in the **Sample pre- and post-deployment script** section at `https://docs.microsoft.com/en-us/azure/data-factory/continuous-integration-deployment#sample-prepostdeployment-script`.

---

15. To configure the production state to deploy the pipeline on the production data factory, switch to the **Pipeline** tab. In the **Stages** section, click on the **Test** stage and then click on the **Copy** icon to clone the **Test** stage:

Figure 8.30 – Cloning the Test stage

16. In the **Stage** pop-up window, enter Production in the **Stage name** field:

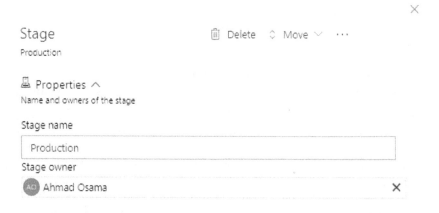

Figure 8.31 – Renaming the cloned stage to Production

Close the **Stage** page. The **Production** stage is configured. We used the environment-name variable when configuring the **Test** stage to make the **ARM template deployment** task generic. Therefore, no changes are required in the **Production** stage.

17. To pass the correct variable values at each stage, select **Variable groups** in the **Variables** tab:

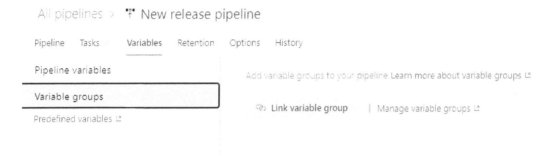

Figure 8.32 – Selecting Variable groups in the Variables tab

18. Click on the **Link variable group** button. In the **Link variable group** page, select the **Test** variable group. In the **Variable group scope** section, select the **Stages** option and then select the **Test** stage:

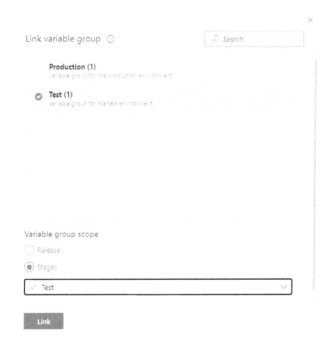

Figure 8.33 – Linking the Test variable group to the Test stage

Click the **Link** button to save the settings.

19. Similarly, link the production variable group to the **Production** stage:

Figure 8.34 – Saving the link settings

Click **Save** to save the pipeline changes.

20. Name the pipeline as `Copy-Data release pipeline`. At this stage, your pipeline is ready to be deployed and should be as shown in the following screenshot:

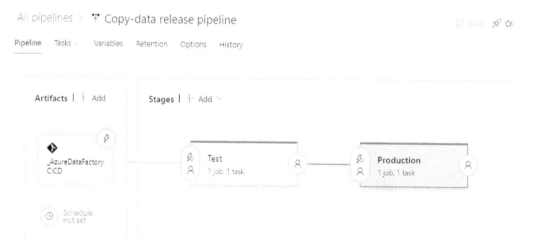

Figure 8.35 – Reviewing the pipeline

We would like the pipeline to deploy automatically to the **Test** environment whenever a new change is published in the Data Factory.

21. To enable continuous deployment, click on the **Continuous deployment** icon on the artifact box:

Figure 8.36 – Enabling continuous deployment

22. On the **Continuous deployment trigger** page, enable the continuous deployment trigger and select the **adf_publish** branch:

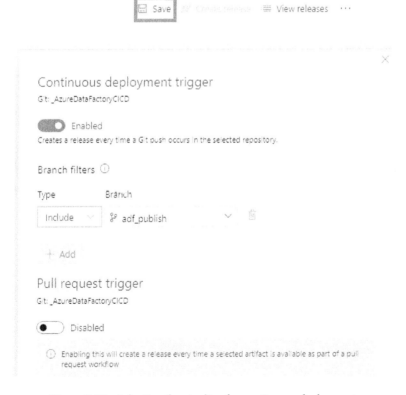

Figure 8.37 – Selecting the pipeline for continuous deployment

A change in the **adf_publish** branch will trigger the deployment. Click **Save** to save the configuration. We can deploy to the **Test** environment as and when changes are available; however, to deploy to the **Production** environment, the changes should be tested and verified. We can do this by adding a manual trigger.

23. To add a manual trigger to deploy to the **Production** stage, click on the **Pre-deployment conditions** button for the **Production** stage:

Figure 8.38 – Adding pre-deployment conditions

24. On the **Pre-deployment conditions** page, select **Manual only** for the trigger:

Figure 8.39 – Selecting the manual trigger

25. Select the approvers in the **Approver** dropdown:

Figure 8.40 – Selecting the approver

Click **Save** to save the pipeline. At this stage, the pipeline should be as shown:

Figure 8.41 – Completing the release pipeline

To see the release pipeline in action, let's switch over to the development data factory in the Azure portal. We'll make a change in the pipeline and will then publish the change to see if it's automatically deployed to the **Test** environment.

26. In the **packt-df-dev** data factory author tab, select **Pipeline-Copy-Data**. We'll add a **Delete** activity to delete the source file once the data is copied to the Azure SQL database. To make changes, we'll create a copy of the **master** branch so that the change will be in a separate branch. To create a new branch, click the down arrow next to the **master** branch in the top menu:

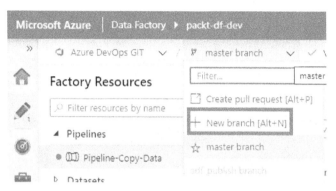

Figure 8.42 – Creating a new branch

27. Name the new branch as `Task123-DeleteSourceFile`. There is no fixed standard syntax with which to name the branch. The name here represents the task ID and a description. The task ID is from a bug management tool such as Jira or the work item ID in Azure DevOps:

Figure 8.43 – Validating the current branch

Observe that the branch name changes to the new value.

28. Drag and drop the **Delete** activity from the **General** tab in the **Activities** section. Name the **Delete** activity as `Delete Source File`. In the **Source** tab of the **Delete** activity, select the `Orders` dataset and leave the rest of the settings as default. In the **Logging** settings, uncheck **Enable logging** to disable the logging:

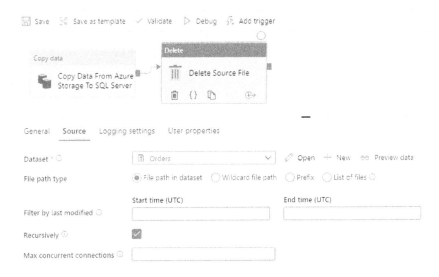

Figure 8.44 – Configuring the Delete activity

Click **Save** to save the changes.

29. Before publishing the changes, click **Debug** to run the pipeline and verify the changes:

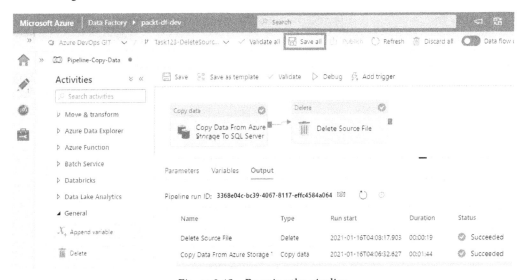

Figure 8.45 – Running the pipeline

The pipeline completes successfully. Click **Save** all to save the changes.

30. To merge the changes from **Task123-DeleteSourceFile** to the **master** branch, click on the drop-down arrow next to **Task123-DeleteSourceFile**:

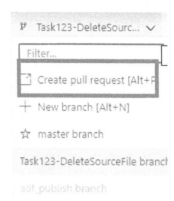

Figure 8.46 – Selecting the Create pull request option

31. Click **Create pull request**. This will open the **New pull request** tab in Azure DevOps. Fill in the **Title** and **Description** fields with `Adding Delete source file feature`. In the **Reviewers** section, add a pull request reviewer. The reviewer will review and approve the code changes to merge them into the **master** branch. You can additionally add the work item and tags to the pull request. You can add yourself as the approver for the purposes of this demo exercise:

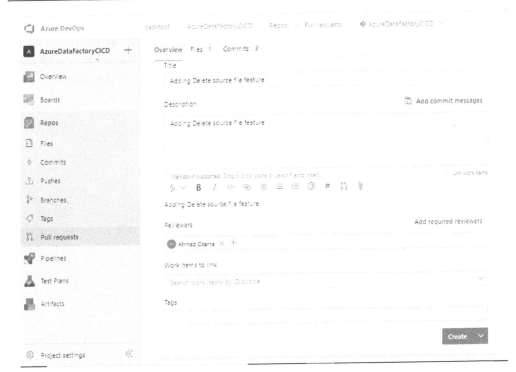

Figure 8.47 – Creating the pull request in Azure DevOps

The pull request is created and the approver is notified via email:

Figure 8.48 – Approving the pull request

32. Click on the **Approve** button to approve the pull request. Click the **Complete** button. In the **Complete pull request** page, select **Merge (no fast forward)** for **Merge type** and leave the rest of the options as their defaults. If you wish to keep the **Test123-DeleteSourceFile** branch, uncheck the **Delete Task123-DeleteSourceFile after merging** option:

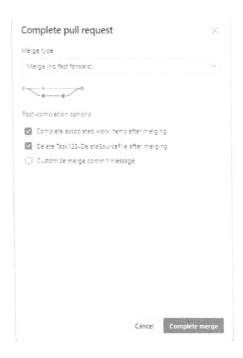

Figure 8.49 – Completing the pull request

Click **Complete merge** to save the changes.

33. Switch to the **packt-df-dev** data factory tab and select **master** for the working branch. Observe that the changes made in the **Task123-DeleteSourceFile** branch are available in the **master** branch:

Figure 8.50 – Publishing the changes to the adf_publish branch

Click **Publish** to publish the changes to the **adf_publish** branch. On the **Pending changes** page, click **OK** to continue.

34. As the changes are published, switch to the **Azure DevOps | Pipelines | Releases** page. Observe that a new release has automatically started:

Figure 8.51 – Releasing changes automatically to the test environment

35. When the deployment completes, switch to the **packt-df-tst** data factory author page in the Azure portal, and then open the **Pipeline-Copy-Data** pipeline:

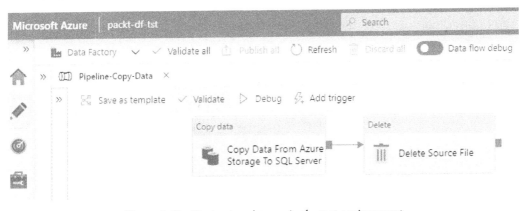

Figure 8.52 – Reviewing changes in the test environment

At this stage, we can test the pipeline for any potential issues and report them to the developers. Click on **Debug** to run and verify the pipeline. When the pipeline completes successfully, switch over to the **Azure DevOps** tab.

36. To deploy the pipeline to production, we'll need to manually approve it. On the **Pipeline | Releases** page, click on the **Production** environment button next to the relevant release number:

Figure 8.53 – Deploying the pipeline to the production environment

37. On the resulting page, click **Deploy**:

Figure 8.54 – Initiating the Production deployment

38. On the **Production Deploy release** page, click **Deploy**:

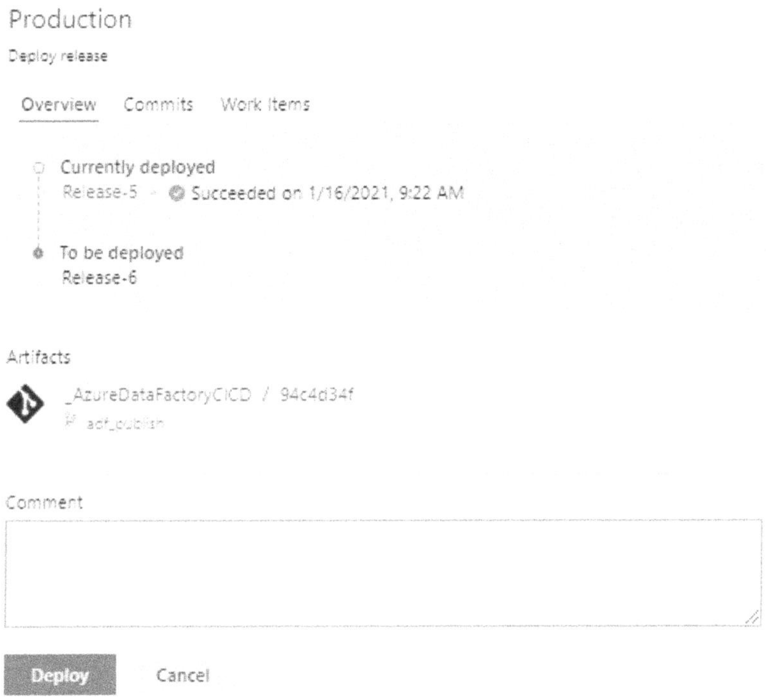

Figure 8.55 – Deploying the pipeline to the production environment

On the resulting page, click **Approve** to start the deployment.

> **Note**
>
> An approval email is sent to the selected approvers with a link to the release to be approved. The approvers can verify and then approve the deployment.

39. In the **Pre-deployment approvals** page, provide any additional comments and then click **Approve** to start the deployment:

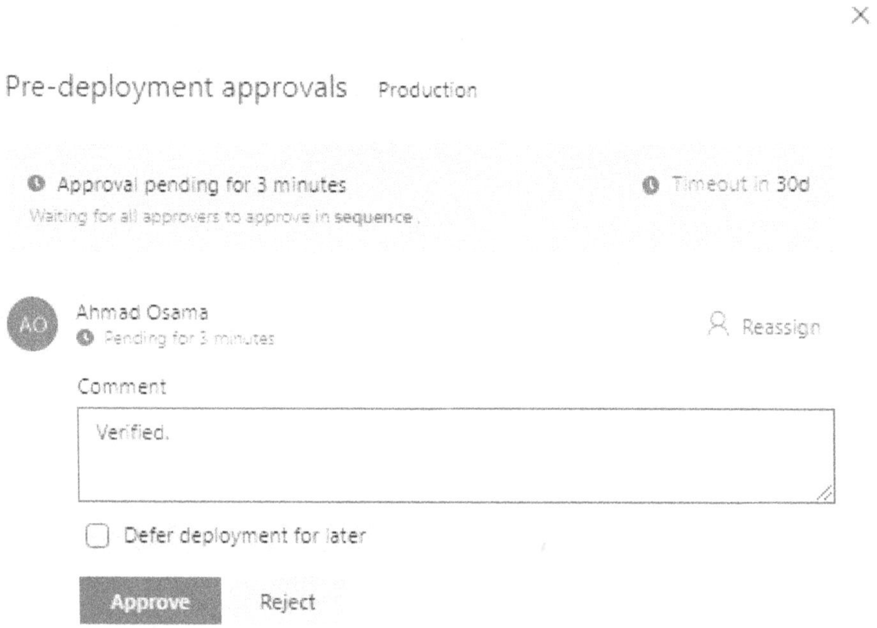

Figure 8.56 – Approving the production deployment

You should get the following output on completion of the deployment:

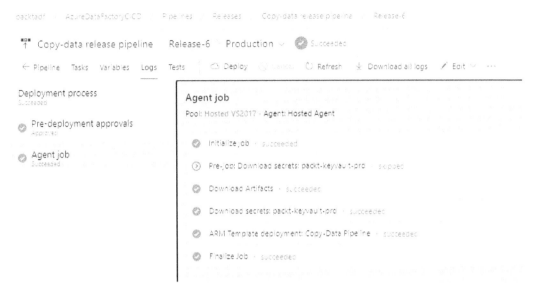

Figure 8.57 – Completing the production deployment

40. When the deployment completes, switch to the **packt-df-prd** data factory in the Azure portal and open the **Pipeline-Copy-Data** pipeline:

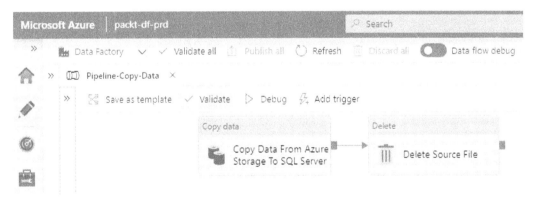

Figure 8.58 – Reviewing the changes in the production environment

This completes the recipe.

9
Batch and Streaming Data Processing with Azure Databricks

Databricks is a data engineering product built on top of Apache Spark and provides a unified, cloud optimized platform so that you can perform ETL, machine learning, and AI tasks on a large quantity of data.

Azure Databricks, as its name suggests, is the Databricks integration with Azure, which further provides fully managed Spark clusters, an interactive workspace for data visualization and exploration, Azure Data Factory, integration with data sources such as Azure Blob Storage, Azure Data Lake Storage, Azure Cosmos DB, Azure SQL Data Warehouse, and more.

Azure Databricks can process data from multiple and diverse data sources, such as SQL or NoSQL, structured or unstructured data, and also scale up as many servers as required to cater to any exponential data growth.

In this chapter, we'll cover the following recipes:

- Configuring the Azure Databricks environment
- Transforming data using Python
- Transforming data using Scala
- Working with Delta Lake
- Processing structured streaming data with Azure Databricks

Let's get started!

Technical requirements

For this chapter, you will need the following

- Microsoft Azure subscription
- PowerShell 7
- Microsoft Azure PowerShell
- Databricks CLI

Configuring the Azure Databricks environment

In this recipe, we'll learn how to configure the Azure Databricks environment by creating an Azure Databricks workspace, cluster, and cluster pools.

Getting ready

To get started, log into `https://portal.azure.com` using your Azure credentials.

How to do it...

An Azure Databricks workspace is the starting point for writing solutions in Azure Databricks. A workspace is where you create clusters, write notebooks, schedule jobs, and manage the Azure Databricks environment.

An Azure Databricks workspace can be created in an Azure managed virtual network or customer managed virtual network. We'll create the environment using a customer managed virtual network.

Creating an Azure Databricks service or workspace

Let's get started with provisioning the virtual network:

1. In Azure portal, type `Virtual Net` into the search box and select **Virtual Networks** from the search results:

Figure 9.1 – Selecting virtual networks

2. On the **Virtual networks** page, click **Add**. On the **Create virtual network** page, under the **Basics** tab, provide **Resource group** with a name, virtual network name, and region:

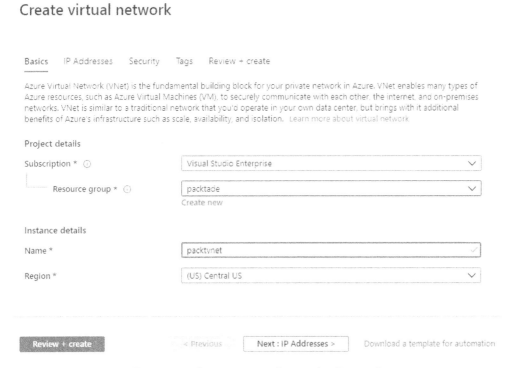

Figure 9.2 – Creating a virtual network – Basics tab

3. Click the **Next: IP Addresses** > button to go to the IP addresses tab. On the IP addresses tab, the IPv4 address space is listed as 10.0.0.0/16 by default. Leave it as-is. Select the default subnet and click **Remove subnet** to delete it:

Figure 9.3 – Removing the default subnet

4. Click **Add subnet**. In the **Add subnet** dialog box, provide a subnet name of databricks-private-subnet and a subnet address range of 10.0.1.0/24. Click **Add** to add the subnet:

Figure 9.4 – Adding a subnet to the virtual network

5. Similarly, add another subnet with a name of `databricks-public-subnet` and an address range of `10.0.2.0/24`:

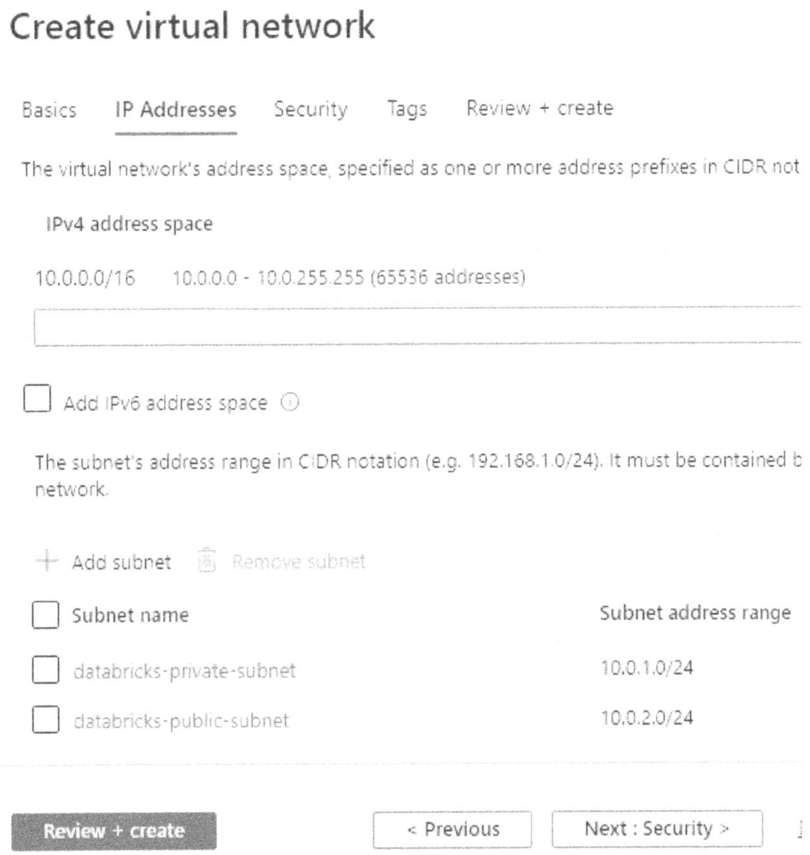

Create virtual network

Basics   **IP Addresses**   Security   Tags   Review + create

The virtual network's address space, specified as one or more address prefixes in CIDR not

IPv4 address space

10.0.0.0/16   10.0.0.0 - 10.0.255.255 (65536 addresses)

☐ Add IPv6 address space ⓘ

The subnet's address range in CIDR notation (e.g. 192.168.1.0/24). It must be contained b network.

╋ Add subnet   🗑 Remove subnet

☐ Subnet name	Subnet address range
☐ databricks-private-subnet	10.0.1.0/24
☐ databricks-public-subnet	10.0.2.0/24

Review + create   < Previous   Next : Security >

Figure 9.5 – Creating a virtual network – IP Addresses

6. Click **Review + create** and then **Create** to create the virtual network. It usually takes 2-5 minutes to create this virtual network.

7. Once the virtual network has been created, click on **Go to Resources** to open the **Virtual network** page. On the **Virtual network** page, select **Subnets** from the **Settings** section:

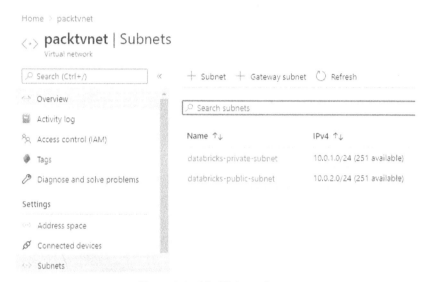

Figure 9.6 – Modifying subnets

8. Click on **databricks-private-subnet**. On the **databricks-private-subnet** page, scroll down to the bottom and select **Microsoft.Databricks/workspaces** from the **Delegate subnet to a service** dropdown:

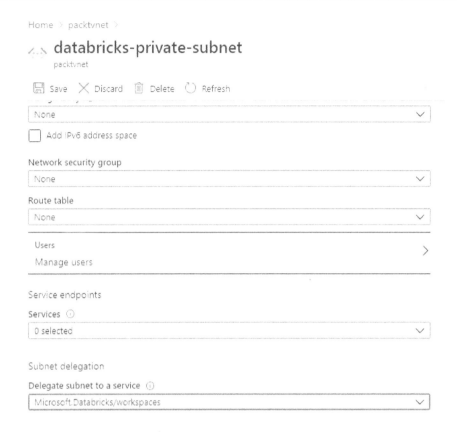

Figure 9.7 – Delegating a subnet to an Azure Databricks workspace

9. Click **Save** to apply the change. Follow the preceding steps to modify `databricks-public-subnet`, similar to how we modified `databricks-private-subnet`. This completes the virtual network and subnet configuration.

10. In the Azure portal, type `Azure Databricks` into the search box and then select **Azure Databricks** from the search list. On the **Azure Databricks** page, click **Add** to create a new Azure Databricks service or workspace.

11. On the **Azure Databricks Service** page, under the **Basics** tab, provide the **Resource group**, **Workspace name**, **Location**, and **Pricing Tier** values (Azure Databricks has two pricing tiers, **Standard** and **Premium**. Premium tier includes all the features of the Standard tier and role-based access. For more details about these tiers, please visit https://azure.microsoft.com/en-in/pricing/details/databricks/.):

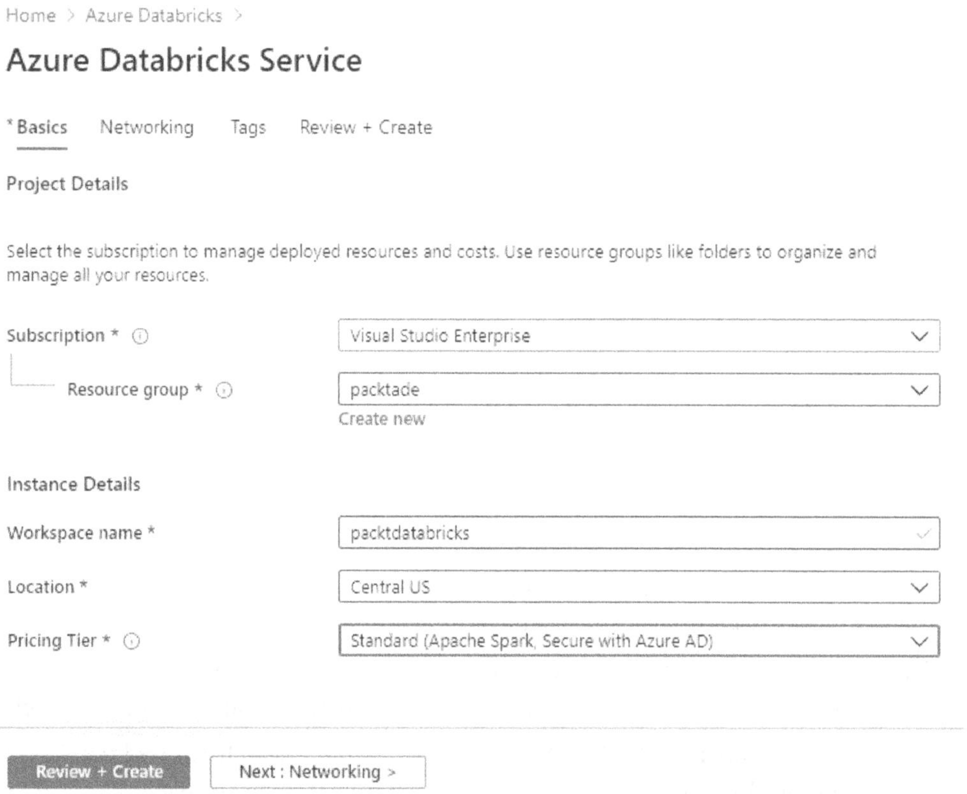

Home > Azure Databricks >

Azure Databricks Service

*Basics   Networking   Tags   Review + Create

Project Details

Select the subscription to manage deployed resources and costs. Use resource groups like folders to organize and manage all your resources.

Subscription * ⓘ | Visual Studio Enterprise   ⌄ |

   └── Resource group * ⓘ | packtade   ⌄ |
                                Create new

Instance Details

Workspace name * | packtdatabricks |

Location * | Central US   ⌄ |

Pricing Tier * ⓘ | Standard (Apache Spark, Secure with Azure AD)   ⌄ |

[Review + Create]   [Next : Networking >]

Figure 9.8 – Creating an Azure Databricks service – Basics tab

12. Click **Next: Networking** > to go to the **Networking** tab. On the **Networking** tab, select **Yes** for **Deploy Azure Databricks workspace in your own Virtual Network (Vnet)**. Select the virtual network we created in *Step 2* from the **Virtual Network** dropdown. Provide the public subnet name, private subnet name, public subnet CIDR range, and private subnet CIDR range that we created in *Step 4*:

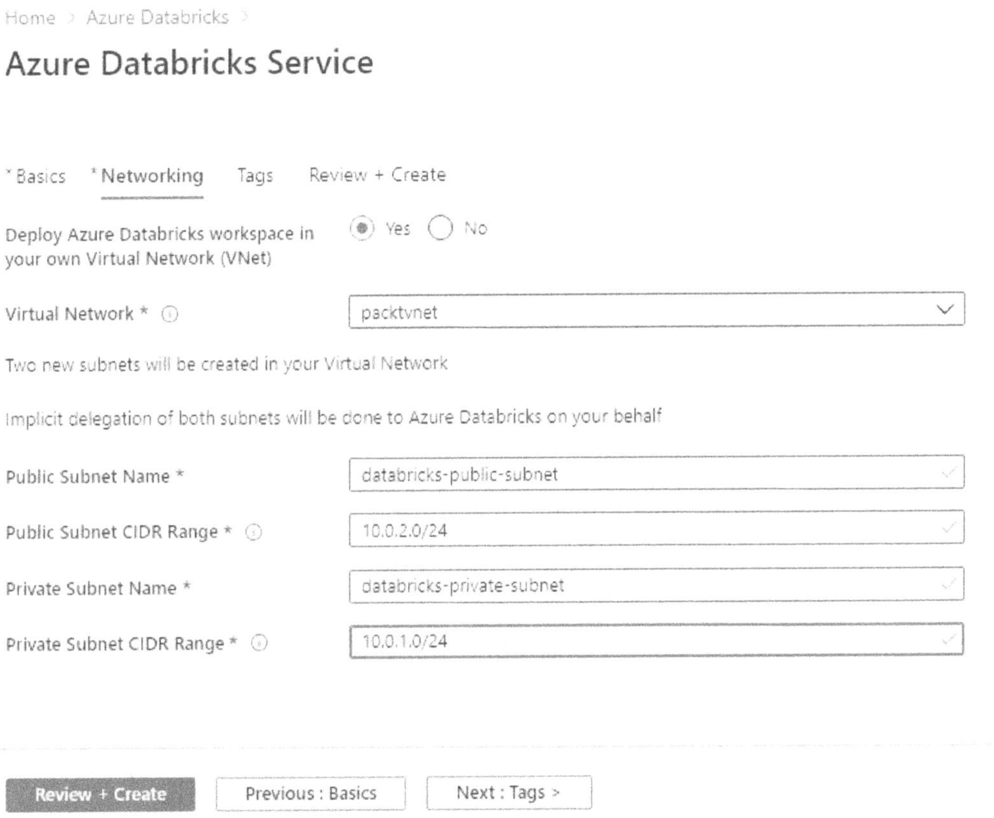

Home > Azure Databricks >

Azure Databricks Service

* Basics   * Networking   Tags   Review + Create

Deploy Azure Databricks workspace in your own Virtual Network (VNet)   ⦿ Yes ◯ No

Virtual Network * ⓘ   packtvnet   ⌄

Two new subnets will be created in your Virtual Network

Implicit delegation of both subnets will be done to Azure Databricks on your behalf

Public Subnet Name *   databricks-public-subnet

Public Subnet CIDR Range * ⓘ   10.0.2.0/24

Private Subnet Name *   databricks-private-subnet

Private Subnet CIDR Range * ⓘ   10.0.1.0/24

Review + Create   Previous : Basics   Next : Tags >

Figure 9.9 – Creating an Azure Databricks service – Networking tab

Azure Databricks uses one public and one private subnet. The public subnet allows you to access the Azure Databricks control plane. Databricks clusters are deployed on the private subnet. It's recommended to not provision any other service in the Databricks private subnet.

> **Note**
> If the public and private subnet don't exist, they'll be created as part of your Azure Databricks workspace automatically.

13. Click **Review + create** and then click **Create** after successful validation. It usually takes around 5-10 minutes to provision a Databricks workspace:

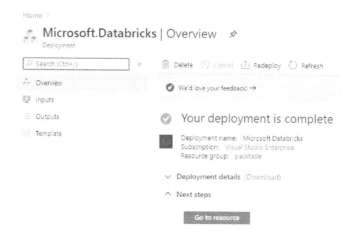

Figure 9.10 – Opening the Azure Databricks workspace

14. Once the service has been created, select **Go to resource** to go to the newly created Azure Databricks workspace:

Figure 9.11 – Azure Databricks service

15. Click on **Launch Workspace** to open the Azure Databricks workspace:

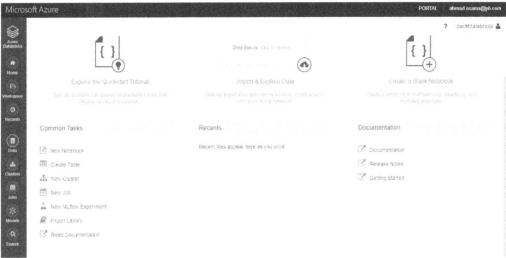

Figure 9.12 – Viewing the Azure Databricks workspace

Databricks authentication is done through Azure **Active Directory** (**AD**). The AD username is displayed at the top-right corner of the workspace page.

An Azure Databricks workspace allows us to create clusters, notebooks, jobs, data sources, and folders so that we can write data transformation or MLFlow experiments. It also helps us organize multiple projects into different folders.

Now, let's create some Azure Databricks clusters.

Creating Azure Databricks clusters

Follow these steps:

1. To create a cluster, select **Clusters** from the left-hand menu of the Databricks workspace:

Figure 9.13 – Opening the Create Cluster page

There are two types of clusters: **Interactive** and **Automated**. Interactive clusters are created manually by users so that they can interactively analyze the data while working on, or even developing, a data engineering solution. Automated clusters are created automatically when a job starts and are terminated as and when the job completes.

2. Click **Create Cluster** to create a new cluster. On the **New Cluster** page, provide
    a cluster name of dbcluster01. Then, set **Cluster Mode** to **Standard**, **Pool**
    to **None**, **Terminate after** to 10 minutes of inactivity, **Min Workers** to 1, **Max
    Workers** to 2, and leave the rest of the options with their defaults:

Create Cluster

| New Cluster | Cancel | Create Cluster | 1-2 Workers: 14.0-28.0 GB Memory, 4-8 Cores, 0.75-1.5 DBU |
| | | | 1 Driver: 14.0 GB Memory, 4 Cores, 0.75 DBU @ |

dbcluster01

Cluster Mode @

Standard

Pool @

None

Databricks Runtime Version @   Learn more

Runtime: 7.0 (Scala 2.12, Spark 3.0.0)

New This Runtime version supports only Python 3

Autopilot Options
☑ Enable autoscaling @
☑ Terminate after   10   minutes of inactivity @

| Worker Type @ | | Min Workers | Max Workers |
| Standard_DS3_v2 | 14.0 GB Memory, 4 Cores, 0.75 DBU | 1 | 2 |

Driver Type

Same as worker   14.0 GB Memory, 4 Cores, 0.75 DBU

▸ Advanced Options

Figure 9.14 – Creating a new cluster

There are two cluster modes: **Standard** and **High Concurrency**. Standard
cluster mode uses single-user clusters, optimized to run tasks one at a time. In a
standard cluster, if there are tasks from multiple users, then a failure in one task
may cause the other task to fail as well. Note that one task can consume all the
cluster's resources, causing another task to wait. High Concurrency cluster mode is
optimized to run multiple tasks in parallel; however, it only supports R, Python, and
SQL workloads and doesn't supports Scala.

These autoscaling options allow Databricks to provision as many clusters as required to process a task within the limit, as specified by the **Min Workers** and **Max Workers** options.

The **Terminate after** option terminates the clusters when there's no activity for a given amount of time. In our case, the cluster will auto terminate after 10 minutes of inactivity. This option helps save costs.

There are two types of cluster nodes: **Worker Type** and **Driver Type**. The driver type node is responsible for maintaining a notebook's state information, interpreting the commands being run from a notebook or a library, and running the Apache Spark master. The worker type nodes are the Spark executor nodes, and these are responsible for distributed data processing.

The **Advanced Options** section can be used to configure Spark configuration parameters, environment variables, tags, configure SSH in the clusters, enable logging, and run custom initialization scripts at the time of cluster creation.

3. Click **Create Cluster** to create the cluster. It will take around 5-10 minutes to create the cluster and may take more time, depending on the number of worker nodes that have been selected:

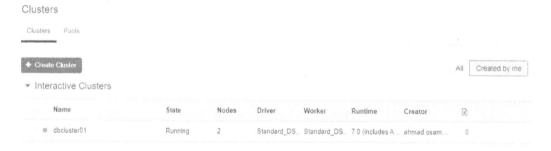

Figure 9.15 – Viewing your clusters

Observe that there are two nodes – one driver and worker – even though the max number of worker nodes is two. Databricks will only create the two worker nodes when the autoscaling condition you've set is reached.

Creating Azure Databricks Pools

Azure Databricks Pools optimize autoscaling by keeping a set of idle, ready-to-use instances without the need for creating instances when required. These idle instances are not charged for. To create Azure Databricks Pools, execute the following steps:

1. In your Azure Databricks workspace, on the **Clusters** page, select the **Pools** page and then select **Create Pool** to create a new pool. Provide the pool's name, then set **Min Idle** to 2, **Max Capacity** to 4, and **Idle Instance Auto Termination** to 10. Leave **Instance Type** as its default of **Standard_DS3_v2** and set **Preloaded Databricks Runtime Version** to **Light 2.4**:

Figure 9.16 – Creating an Azure Databricks Pool

Min Idle specifies the number of instances that will be kept idle and available and won't terminate. The **Idle Instance Auto Terminate** settings doesn't apply to these instances.

Max Capacity limits the maximum number of instances to this number, including idle and running ones. This helps with managing cloud quotas and their costs. The Azure Databricks runtime is a set of core components or software that run on your clusters. There are different runtimes, depending on the type of workload you have. To find out more about the Azure Databricks runtime, please visit https://docs.microsoft.com/en-us/azure/databricks/runtime/.

2. Click **Create** to create the pool. We can attach a new of existing or cluster to a pool by specifying the pool name under the **Pool** option. In the workspace, navigate to the **Clusters** page and select **dbcluster01**, which we created in *Step 2* of the previous section. On the **dbcluster01** page, click **Edit** and select **dbclusterpool** from the **Pool** drop-down list:

Figure 9.17 – Attaching an existing cluster to a pool

3. Click **Confirm** to apply these changes. The cluster will now show up in the **Attached Clusters** list:

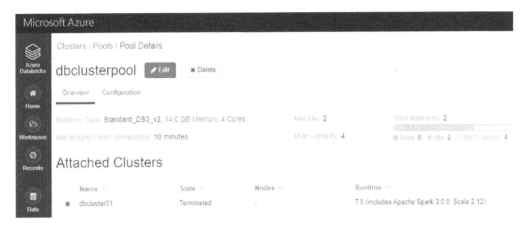

Figure 9.18 – Viewing clusters that have been attached to a pool

We can add multiple clusters to a pool; however, we should modify the number of idle clusters and their maximum instance capacity accordingly.

Whenever an instance, such as dbcluster01, requires an instance, it'll attempt to allocate the pool's idle instance. If an idle instance isn't available, the pool expands to provision new instances.

Transforming data using Python

Data transformation at scale is one of the most important uses of Azure Databricks. In this recipe, we'll read product orders from an Azure storage account, read customer information from an Azure SQL Database, join the orders and customer information, apply transformations to filter and aggregate the total order by country and customers, and then insert the output into an Azure SQL Database.

Getting ready

To get started, follow these steps:

1. Log into https://portal.azure.com using your Azure credentials.

2. You will need an existing Azure Databricks workspace and at least one Databricks cluster. You can create these by following the *Configuring an Azure Databricks environment* recipe.

How to do it...

Let's get started by creating an Azure storage account and an Azure SQL database:

1. Execute the following command to create an Azure Storage account and upload the orders files to the orders/datain container:

```
.\azure-data-engineering-cookbook\Chapter06\2_
UploadDatatoAzureStorage.ps1 -resourcegroupname
packtade -storageaccountname packtstorage -location
centralus -datadirectory C:\azure-data-engineering-
cookbook\Chapter09\Data\  -createstorageaccount $false
-uploadfiles $true
```

The preceding command creates an Azure Storage account called packtstorage, creates a container called orders, uploads the files in the Data folder, and outputs the storage account key. Copy and save the key1 value so that it can be used later.

2. Execute the following command to create an Azure SQL Database:

```
.\azure-data-engineering-cookbook\Chapter06\1_
ProvisioningAzureSQLDB.ps1 -resourcegroup packtade
-servername azadesqlserver -databasename azadesqldb
-password Sql@Server@1234 -location centralus
```

The preceding command will create an Azure SQL Server called azadesqlserver and a database called azadesqldb with a Basic performance tier.

3. Execute the following script to create the required tables in the Azure SQL database:

```
DROP TABLE IF EXISTS Customer
GO
CREATE TABLE Customer
  (
      id INT,
      [name] VARCHAR(100)
  )
GO
INSERT INTO customer
  VALUES(1,'Hyphen'),(2,'Page'),(3,'Data
Inc'),(4,'Delta'), (5,'Genx'),(6,'Rand Inc'),(7,'Hallo
Inc')
```

```
DROP TABLE IF EXISTS SalesStaging
GO
CREATE TABLE SalesStaging
  (
      CustomerName VARCHAR(100),
      Country VARCHAR(100),
      Amount DECIMAL(10,2)
  )
```

4. To connect to our Azure SQL Database from Azure Databricks, we'll need to add a service endpoint to `databricks-public-subnet`. We can also connect to our Azure SQL Database from Azure Databricks by enabling the firewall to *allow Azure services and resources to access this server* in our Azure SQL Database; however, that's not as secure as connecting via a service endpoint. To create a service endpoint, in the Azure portal, open our Azure SQL Database and navigate to **Firewall settings**:

Figure 9.19 – Opening the Azure SQL Database Firewall settings

5. On the **Firewall settings** page, scroll down to **Virtual networks** and select **+ Add existing virtual network**.

6. On the **Create/Update** virtual network rule, provide the rule name, subscription, and the virtual network. Under **Subnet name/ Address prefix**, provide the Databricks public subnet (`databricks-public-subnet`):

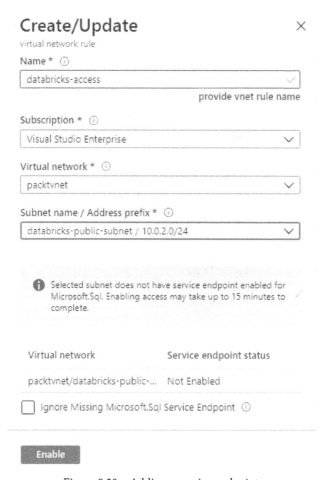

Figure 9.20 – Adding a service endpoint

7. Click **Enable** and then click **OK** to create the endpoint.

8. Now, we'll create some Databricks secrets so that we can store the Azure Storage account key and Azure SQL Database login credentials. Secrets allow us to fetch secure information without adding it as plain text to the notebook. These secrets can be stored in Azure Key Vault or Databricks. We'll use Databricks. To do this, open a command prompt and execute the following command to configure the Databricks CLI:

```
databricks configure --token
```

You'll be asked to provide the Databricks host and token. To generate the token, switch to our Databricks workspace in the Azure portal and then click the username button in the top-right corner:

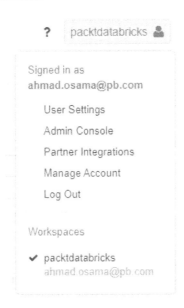

Figure 9.21 – Getting the Databricks access token

9. Select **User Settings** and then select **Generate New Token** under the **Access Tokens** tab. Copy and paste the token ID from the command prompt. To get the Databricks host, open the Azure Databricks Service, and then copy the URL:

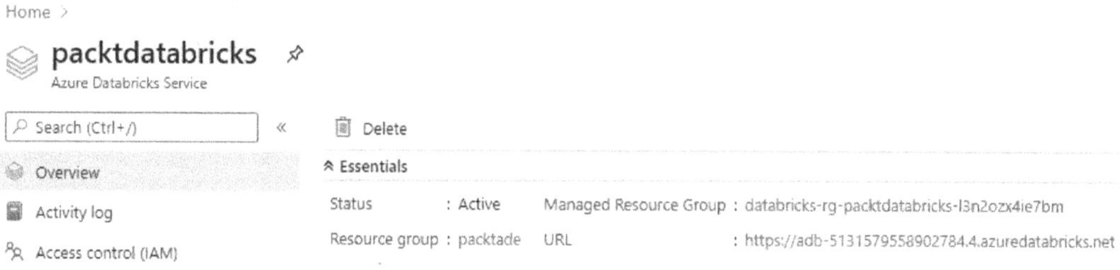

Home >

packtdatabricks ⚲
Azure Databricks Service

🔍 Search (Ctrl+/) «	🗑 Delete
🔷 Overview	⊼ Essentials
📋 Activity log	Status : Active Managed Resource Group : databricks-rg-packtdatabricks-l3n2ozx4ie7bm
🔏 Access control (IAM)	Resource group : packtade URL : https://adb-5131579558902784.4.azuredatabricks.net

Figure 9.22 – Getting the Databricks host

10. Once the configuration is completed, execute the following command:

```
databricks secrets create-scope --scope azuresql
--initial-manage-principal users
```

The preceding code creates a scope called `azuresql` for storing the Azure SQL Database's credentials.

11. Execute the following command to add an Azure SQL username to the `azuresql` secret scope:

```
databricks secrets put --scope azuresql -key username
```

A notepad will appear. Type the username value into the notepad, and then save and close it.

12. Execute the following command to store the Azure SQL Database's password:

```
databricks secrets put --scope azuresql --key password
```

13. Similarly, execute the following command to create the `azurestorage` secret scope and store the Azure storage account key:

```
databricks secrets create-scope --scope azurestorage
--initial-manage-principal users
databricks secrets put --scope azurestorage --key
accountkey
```

14. Now, let's create a Python notebook so that we can connect to and transform the data. To do that, open the Azure Databricks service and select **Launch Workspace** to log into your Azure Databricks workspace.

> **Note**
> If you don't have time to follow along, you can import the ~/Chapter09/
> TransformDataUsingPython.dbc notebook. To import this
> notebook, right-click anywhere in the **Workspace** tab and then select **Import**.
> Browse or drag and drop the notebook and click **Import**.

In your Azure Databricks workspace, from the left menu bar, select **Workspace**:

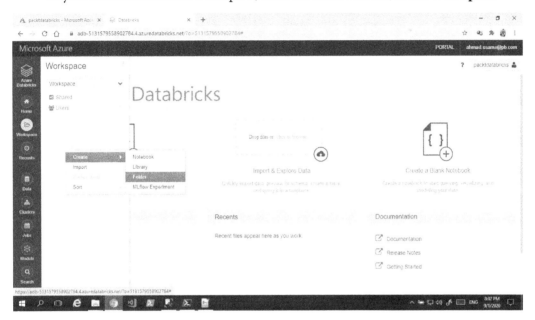

Figure 9.23 – Creating a folder in your Azure Databricks workspace

Right-click anywhere in the **Workspace** tab, select **Create**, and then select **Folder**.
Use Demos as the folder's name and select **Create folder**.

15. On the **Workspace** tab, right-click **Demos**, select **Create**, and then select **Notebook**:

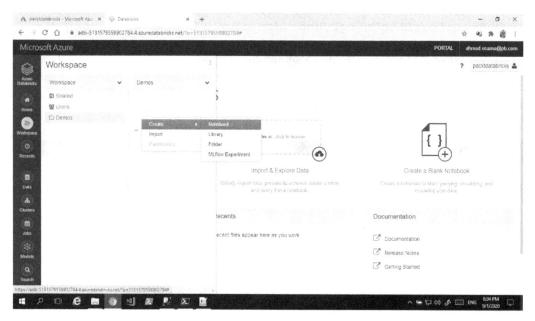

Figure 9.24 – Creating a new notebook

> **Note**
> You will need a running cluster to create a notebook. If the cluster is in a terminated state, start the cluster from the **Cluster** page in the left menu bar.

16. In the **Create Notebook** dialogue box, provide a notebook name of
TransformDataUsingPython, set **Default Language** to **Python**, and select the cluster:

Figure 9.25 – Providing notebook parameters

> **Note**
>
> We can switch between languages in a notebook; however, we do need to provide a default language.

17. Click **Create** to create the notebook. In the new notebook, copy and paste the following code to add a title and description to the notebook:

```
%md # Transform data using Python
%md Data transformation at scale is one of the most
important use of Azure Databricks. In this recipe, we'll
read product orders from an Azure storage account, read
customer information from Azure SQL Database, join the
orders and customer information, apply transformation to
filter and aggregate total order by country and customers
and then insert the output into an Azure SQL Database.
We'll then schedule the job to run automatically.
```

Commands starting with % are referred to as **magic commands**. The %md # line specifies the title, while the %md line is for adding text. Your notebook will look as follows:

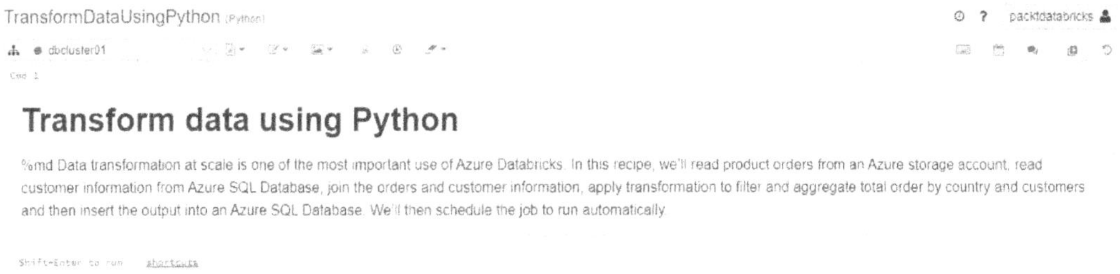

TransformDataUsingPython (Python)   ⊘ ?   packtdatabricks ▲

⊿ ● dbcluster01

Transform data using Python

%md Data transformation at scale is one of the most important use of Azure Databricks. In this recipe, we'll read product orders from an Azure storage account, read customer information from Azure SQL Database, join the orders and customer information, apply transformation to filter and aggregate total order by country and customers and then insert the output into an Azure SQL Database. We'll then schedule the job to run automatically.

Figure 9.26 – Adding a description and a heading to your notebook

18. Press *B* to add a new cell to the notebook. Copy and paste the following code into the new cell:

```
# specify credentials using secrets
accountkey = dbutils.secrets.get(scope = "azurestorage",
                                  key = "accountkey")
# specify blob storage name and account key
spark.conf.set(
  "fs.azure.account.key.packtstorage.blob.core.windows.
net",
  accountkey)
```

```
# specify the schema for the orders file
# specify the schema as the orders file don't have
headers.
# if the file has headers defined you can use inferSchema
option to import the schema
from pyspark.sql.types import *
orderschema = StructType([
  StructField("invoiceno",StringType(),True),
  StructField("stockcode",StringType(),True),
  StructField("description",StringType(),True),
  StructField("quantity",IntegerType(),True),
  StructField("invoicedate",StringType(),True),
  StructField("unitprice",DoubleType(),True),
  StructField("customerid",StringType(),True),
  StructField("Country",StringType(),True),
  StructField("orderid",IntegerType(),True)])


# read the orders files from the datain folder in orders
container
ordersdf = spark.read.csv(
    "wasbs://orders@packtstorage.blob.core.windows.net/
datain/",
    schema= orderschema, # specify the schema defined
above
    sep="|") # specify the separator
display(ordersdf) # display the dataframe
```

> **Note**
>
> It is easy to work with shortcuts in a notebook. To get a list of shortcuts and magic commands, please visit https://docs.databricks.com/notebooks/notebooks-use.html.

The preceding code gets the Azure Storage account key value from the secrets using dbutils.secrets.get. It then defines the orders file schema using the StructType command and reads the orders1.txt files using the spark.read.csv command. The spark.read.csv file specifies the Azure storage folder, the schema, and the separator, and then reads the file's content into a DataFrame. A DataFrame is an in-memory SQL-like table data structure that makes it easy to process data, as we'll see in this recipe.

> **Note**
>
> The orders1.txt file doesn't have column names, so the schema is specified explicitly. If the file contains headers, we can use the inferSchema option in the spark.read.csv file to get the column's name and schema.

19. We can also add a title from the UI. To do that, select the down arrow from the top-right corner of the cell and select **Show title**:

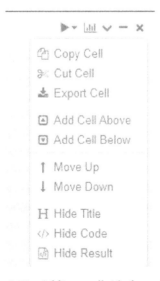

Figure 9.27 – Adding a cell title from the UI

We can also use the preceding menu to hide and show results, hide and show code, and cut/copy and add new cells.

20. Set the title to `Read data from Azure Blob Storage`. We can run all or specific cells in a notebook. Hit *Shift + Enter* to run the cell. You should get the following output:

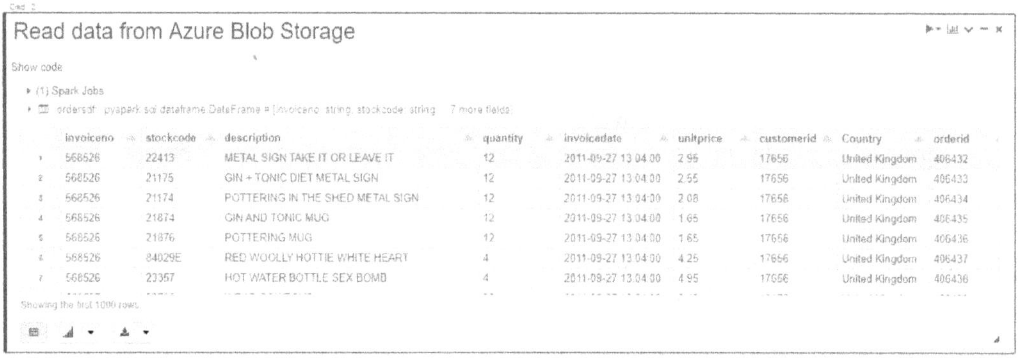

Figure 9.28 – Reading data from Azure Storage

> **Note**
> The relevant code was hidden when we took the preceding screenshot for the sake of brevity.

Observe that the data is read and that the schema is applied to the file. The file data is in a DataFrame, and we can use it to perform transformations. The result is displayed in a tabular format by default. However, we can also use multiple graphs to visualize the data by grouping and aggregating on various data points. We can do this by clicking on the bar chart icon, shown in the bottom-left corner of the preceding screenshot.

21. Let's connect to our Azure SQL Database and get the customer data. To do that, add a new cell and then copy and paste the following code into the cell. Add our `Read Customer Information from Azure SQL Database` title to the cell:

```
#Customer information is stored in Customer table in an
Azure SQL Database
# Azure SQL Server FQDN
jdbcHostname = "aadesqlserver.database.windows.net"
jdbcDatabase = "azadesqldb" # Azure SQL Database name
jdbcPort = 1433 # SQL Server default port
# get username and password for Azure SQL Database from
```

```
database secrets
username = dbutils.secrets.get(scope = "azuresql", key =
"username")
password = dbutils.secrets.get(scope = "azuresql", key =
"password")
# set the JDBC URL
jdbcUrl = \
"jdbc:sqlserver://
{0}:{1};database={2};user={3};password={4}".format(
    jdbcHostname, jdbcPort, jdbcDatabase, username,
password)
# read customer table.
customerdf = spark.read.jdbc(url=jdbcUrl,
table="customer")
display(customerdf)
```

The preceding code gets the Azure SQL Database credentials from the Databricks secrets, and then uses a JDBC driver to connect to our Azure SQL Database and fetch and place the customer table in the `customerdf` DataFrame.

22. Press *Shift + Enter* to run the cell. You should get the following output:

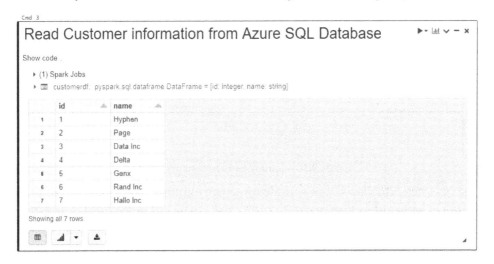

Figure 9.29 – Reading the Customer table from our Azure SQL Database

23. Now, we'll join the `ordersdf` and `customersdf` information to get the customer's name. To do that, add a new cell and copy and paste the following code. Give the cell a title; in this case, `Join Orders and Customer Data`:

```
# join ordersdf and customerdf and select only the
required columns
customerordersdf = ordersdf.join(customerdf,
ordersdf.customerid == customerdf.id, how="inner").
select(ordersdf.Country, ordersdf.unitprice, ordersdf.
quantity, customerdf.name.alias("customername") )
display(customerordersdf)
```

The preceding code joins `ordersdf` and `customerdf` on `ordersdf.customerid` and `customerdf.id`, respectively, and only selects the required columns (for further calculations) in the `customerordersdf` DataFrame.

You should get the following output:

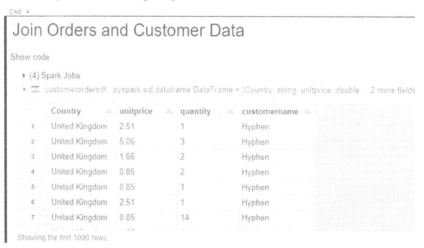

Figure 9.30 – Joining two DataFrames

24. The next step is to filter out or exclude rows with a country name of `unspecified` whose quantity is less than zero. To do this, add a new cell called `Filter out irrelevant data` and copy and paste the following code into the new cell:

```
from pyspark.sql.functions import col
filterdf = customerordersdf.
filter((col("country")!="Unspecified") &
(col("quantity")>0)).withColumn("totalamount",
col("unitprice")*col("quantity"))
display(filterdf)
```

The preceding code uses the `filter` command to exclude rows with country set to `Unspecified` and with a quantity less than zero. It also uses the `withColumn` command to add a new column, `totalamount`, to the DataFrame. The total amount is calculated as *unitprice * quantity* for each row.

Press *Shift + Enter* to run the cell. You should get the following output:

Figure 9.31 – Filtering and calculating totalamount

Note that we can combine *step 23* and *step 24* into a single command, as shown here:

```
from pyspark.sql.functions import col
customerordersdf = ordersdf.join(customerdf,
ordersdf.customerid == customerdf.id, how="inner").
select(ordersdf.Country, ordersdf.unitprice, ordersdf.
quantity, customerdf.name.alias("customername")
).filter((col("country")!="Unspecified") &
(col("quantity")>0)).withColumn("totalamount",
col("unitprice")*col("quantity"))
```

25. Now, we'll calculate the total sales by customer and country. To do this, copy and paste the following code into a new cell. Name the cell `Calculate total sales by Customer and Country`:

```
import pyspark.sql.functions as fun
salesbycountries = filterdf.groupBy("customername",
"country").agg(fun.sum("totalamount").alias("Amount"))
display(salesbycountries)
```

The preceding code uses the groupBy clause to group the data by customer name and country, and then adds the total amount for each group to get the total sales. You should get the following output:

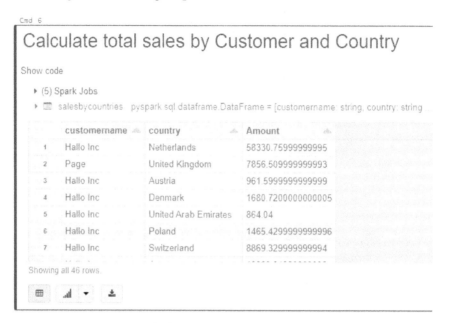

Figure 9.32 – Calculate total sales by Customer and Country

26. Now, we'll insert the salesbycountries DataFrame from the previous step into our Azure SQL Database. To do that, in a new cell, copy and paste the following code. Call the cell Insert sales in Azure SQL Database:

```
# insert data into SalesStaging table.
# mode can be append to add new rows or overwrite to
truncate and insert the data.
salesbycountries.write.jdbc(jdbcUrl,"dbo.
SalesStaging",mode="overwrite")
# read the SalesStaging table.
from pyspark.sql.functions import desc
salesstagingdf = spark.read.jdbc(url=jdbcUrl,
table="SalesStaging")
display(salesstagingdf.sort(desc("Amount")))
```

The preceding command uses dataframe.write.jdbc to insert the DataFrame into the SalesStaging table in our Azure SQL Database. It then reads the table and displays the values in descending order of amount.

> **Note**
>
> For more information on Python DataFrame commands, visit `https://docs.databricks.com/spark/latest/dataframes-datasets/introduction-to-dataframes-python.html`.

Transforming data using Scala

In this recipe, we'll mount the Azure Data Lake Storage Gen2 filesystem on DBFS. We'll then read the `orders` data from Data Lake and the `customer` data from an Azure Synapse SQL pool. We'll apply transformation using Scala, analyze data using SQL, and then insert the aggregated data into an Azure Synapse SQL pool.

Getting ready

To get started, follow these steps:

1. Log into `https://portal.azure.com` using your Azure credentials.

2. You will need an existing Azure Databricks workspace and at least one Databricks cluster. You can create these by following the *Configuring an Azure Databricks environment* recipe.

How to do it...

Let's get started with provisioning the source and the destination data sources. We'll begin by creating and uploading files to an Azure Data Lake Storage Gen2 account:

1. Execute the following command to create an Azure Data Lake Storage Gen2 account and upload the necessary files:

```
.\azure-data-engineering-cookbook\Chapter04\1_
UploadOrderstoDataLake.ps1 -resourcegroupname packtade
-storageaccountname packtdatalake -location centralus
-directory C:\azure-data-engineering-cookbook\Chapter04\
Data\
```

2. We'll access the Azure Data Lake Storage Gen2 account from Azure Databricks using a service principal. Execute the following command to create a service principal:

```
.\azure-data-engineering-cookbook\Chapter05\2_
CreateServicePrincipal.ps1 -resourcegroupname packtade
-password Awesome@1234 -name databricks-service-principal
```

Take note of the application ID and ID from the output of the preceding command.

3. Now, we'll create an Azure Synapse SQL pool. To do that, execute the following PowerShell command:

```
.\azure-data-engineering-cookbook\Chapter03\1_
ProvisioningAzureSynapseSqlPool.ps1
```

The preceding command creates an Azure SQL Server called azadesqlserver and the SQL Pool, which will be called azadesqldw. The username and password are sqladmin and Sql@Server@1234, respectively. If you wish to change the database's name, username, or password, you can do so by modifying the parameters in the 1_ProvisioningAzureSynapseSqlPool.ps1 script. When the SQL pool is created and is available, run the SQL queries in the ~/Chapter09/Databaseobjects.sql file to create the Customer and SalesStaging tables in the SQL Pool. You'll also need to add the virtual network firewall rule to allow connections from the Databricks public subnet, as we did in the previous recipe.

> **Note**
>
> If you don't have time to follow along, you can import the ~/Chapter09/TransformDataUsingScala.dbc notebook. To import this notebook, right-click anywhere in the **Workspace** tab and select **Import**. Browse or drag and drop the notebook and click **Import**.

4. Now, we'll create the notebook. In the Azure portal, open the Databricks service and click **Launch Workspace**.

5. In the Databricks workspace, from the left bar, select **Workspace** and then create a new notebook under the **Demo** folder. Name the notebook TransformingDataUsingScalaAndSQL, set **Default Language** to **Scala**, select the cluster, and create the notebook.

Unlike the previous recipe, instead of accessing Azure Storage directly, we'll mount the storage onto the DBFS. To do that, copy and paste the following code into the notebook:

```
// mount Azure Data Lake Storage file system (container)
to databricks file system.
// configure credentials
val configs= Map(
    "fs.azure.account.auth.type" -> "OAuth",
    "fs.azure.account.oauth.provider.type" -> "org.apache.
hadoop.fs.azurebfs.oauth2.ClientCredsTokenProvider",
    "fs.azure.account.oauth2.client.id" -> "fcadb570-4a04-
4776-b8f9-6725cab5adde",
    "fs.azure.account.oauth2.client.secret" -> dbutils.
secrets.get(scope = "databricksserviceprincipal", key =
"password"),
    "fs.azure.account.oauth2.client.endpoint" -> "https://
login.microsoftonline.com/8a4925a9-fd8e-4866-b31c-
f719fb05dce6/oauth2/token")


// mount the file system
dbutils.fs.mount(
    source = "abfss://ecommerce@packtdatalake.dfs.core.
windows.net/",
    mountPoint = "/mnt/ecommerce",
    extraConfigs = configs)
```

In the preceding code, replace fs.azure.account.oauth2.client.id with the service principal application ID from *Step 2*. Replace the uniqueidentifier value (8a4925a9-fd8e-4866-b31c-f719fb05dce6) in fs.azure. account.oauth2.client.endpoing with your Azure tenant ID. Provide the service principal password from *Step 2* in fs.azure.account.oauth2. client.secret. You can either provide the password in plain text or you can use a Databricks secret. Replace ecommerce with the filesystem's name or container name and packtdatalake with the Azure Storage account name.

6. Hit *Shift + Enter* to run the cell. The ecommerce filesystem will be mounted on the /mnt/ecommerce directory on the DBFS. To verify this, execute the following command in a new cell to list the files in the ecommerce filesystem:

```
display( dbutils.fs.ls("/mnt/ecommerce/orders"))
```

You should get the following output:

```
1  display( dbutils.fs.ls("/mnt/ecommerce/orders"))
```

	path	name	size
1	dbfs:/mnt/ecommerce/orders/orders1.txt	orders1.txt	12601902
2	dbfs:/mnt/ecommerce/orders/orders2.tif	orders2.tif	12758599
3	dbfs:/mnt/ecommerce/orders/orders2.txt	orders2.txt	12735799
4	dbfs:/mnt/ecommerce/orders/orders3.tif	orders3.tif	12758599
5	dbfs:/mnt/ecommerce/orders/orders3.txt	orders3.txt	12758599
6	dbfs:/mnt/ecommerce/orders/orders4.tif	orders4.tif	12758599
7	dbfs:/mnt/ecommerce/orders/orders4.txt	orders4.txt	12886718

Showing all 11 rows.

Figure 9.33 – Listing files in the Azure Data Lake Container

7. Now, we'll read the `orders` data from the data lake. To do that, copy and paste the following code into a new cell:

```
import org.apache.spark.sql.types._

// specify the orders file schema.
// The orders text file don't have headers. If the file
has headers, we can use
// inferSchema to get column datatypes
val orderSchema = StructType(Array(
  StructField("invoiceno", StringType, true),
  StructField("stockcode", StringType, true),
  StructField("description", StringType, true),
  StructField("quantity", IntegerType, true),
  StructField("invoicedate", StringType, true),
  StructField("unitprice", DoubleType, true),
  StructField("customerid", IntegerType, true),
  StructField("Country", StringType, true),
  StructField("orderid", IntegerType, true))
)
// read all the text files in the orders container.
```

```
val orders = spark.read
.format("com.databricks.spark.csv")
.option("header","true")
.option("delimiter","|")
.schema(orderSchema)
.load("/mnt/datalake/orders/")

// display 1000 records
display(orders)
```

The preceding command defines the text file schema and reads all the text files in the `orders` container into the `orders` DataFrame.

8. Now, we'll fetch the customer information from the Azure Synapse SQL Pool. To connect to the Synapse SQL Pool, we'll use a Spark connector instead of JDBC. To use a Spark connector, we'll have to add the `spark-connector` jar library to the Databricks cluster. To do that, in the Azure Databricks workspace, select **Workspace** from the right bar, right-click anywhere on the **Workspace** tab, select **Create**, and then select **Library**. The Databricks cluster should be up and running so that you can upload the `.jar` file:

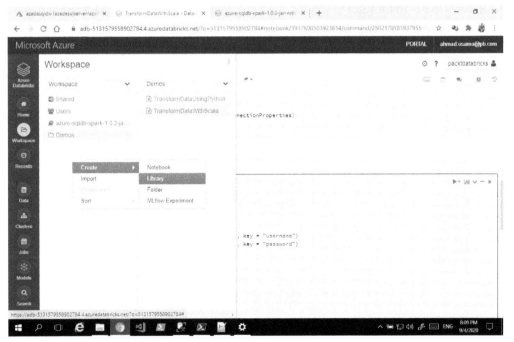

Figure 9.34 – Creating a new library

9. In the **Create Library** window, select the **Upload** tab and then drag and drop the `C:\azure-data-engineering-cookbook\Chapter09\azure-sqldb-spark-1.0.2-jar-with-dependencies.jar` file onto the library page:

Figure 9.35 – Uploading the .jar to the library

10. It may take 5-10 minutes to upload the file, depending on your network speed. Click **Create** to go to the next step:

Figure 9.36 – Installing the .jar on the cluster

11. Select the cluster and then click **Install** to install the jar on the selected cluster. You may have to restart the cluster once it's been installed. Copy and paste the following code into a new notebook cell:

```
import com.microsoft.azure.sqldb.spark.config.Config
import com.microsoft.azure.sqldb.spark.connect._

val hostname = "azadesqlserver.database.windows.net"
val database = "azadesqldw"
val port = 1433
val username = dbutils.secrets.get(scope = "azuresql",
key = "username")
val password = dbutils.secrets.get(scope = "azuresql",
key = "password")

val config = Config(Map(
  "url"            -> hostname,
  "databaseName"   -> database,
  "dbTable"        -> "dbo.customer",
  "user"           -> username,
  "password"       -> password,
  "connectTimeout" -> "5", //seconds
  "queryTimeout"   -> "5" //seconds
))

val customer = sqlContext.read.sqlDB(config)
display(customer)
```

The preceding code connects to the Synapse SQL Pool and puts the data from the customer table into the customer DataFrame.

Now, let's aggregating data on quantity, unitprice, country, and customer. Let's explore and clean the data if required.

12. The quantity and untiprice columns should be any value greater than zero. To analyze the two columns, copy and paste the following code into a new cell:

```
display(orders.describe( "quantity","unitprice"))
```

The preceding code returns the summary of the column, as shown here:

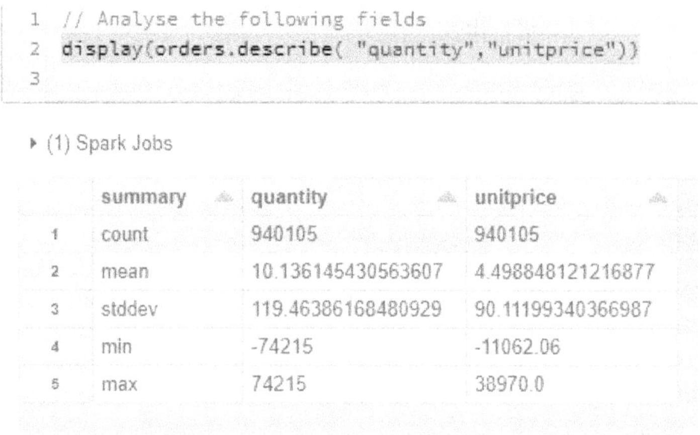

```
1  // Analyse the following fields
2  display(orders.describe( "quantity","unitprice"))
3
```

▶ (1) Spark Jobs

	summary	quantity	unitprice
1	count	940105	940105
2	mean	10.136145430563607	4.498848121216877
3	stddev	119.46386168480929	90.11199340366987
4	min	-74215	-11062.06
5	max	74215	38970.0

Figure 9.37 – Describing column values

13. Observe that we have negative values for quantity and unitprice. Copy and paste the following code to filter out the negative values:

```
val cleanOrders = orders.where($"quantity">"0" &&
$"unitprice">"0")
display(cleanOrders.describe( "quantity", "unitprice"))
```

The preceding filters out values greater than zero for the quantity and unitprice columns.

14. Let's analyze the values in the country column. To do that, copy and paste the following code into a new cell:

```
import org.apache.spark.sql.functions._
// Find the distinct last names for each first name
display(cleanOrders.select("country").distinct.
sort("country"))
```

The preceding commands gets the distinct country names in ascending order. You should get the following output:

Figure 9.38 – Getting distinct country names

15. Observe that there's Unspecified as a country name. Execute the following command to replace Unspecified with India:

```
import org.apache.spark.sql.functions._
// Find the distinct last names for each first name
//display(cleanOrders.select("country").distinct.
sort("country"))
val cleanCountry = cleanOrders.withColumn("country",
when(col("country")=== "Unspecified", "India").
otherwise(col("country")))
display(cleanCountry.select("country").distinct.
sort("country"))
```

The preceding code uses the when command to replace the country value of Unspecified with India.

16. Now, let's remove any duplicates. Copy and paste the following command into a new cell:

```
// Get total rows
print("Total Rows", cleanCountry.count)
// remove duplicates. We can also use --
// val distinctOrders = cleanCountry.dropDuplicates() --
to remove duplicate based on column
// val countOrders = cleanCountry.
```

```
dropDuplicates("orderid","invoicedate")  -- to remove
duplicate based on column/s
val distinctOrders = cleanCountry.distinct()
print("Distinct Rows", distinctOrders.count)
```

The preceding command removes the duplicate rows from the DataFrame. You should get the following output:

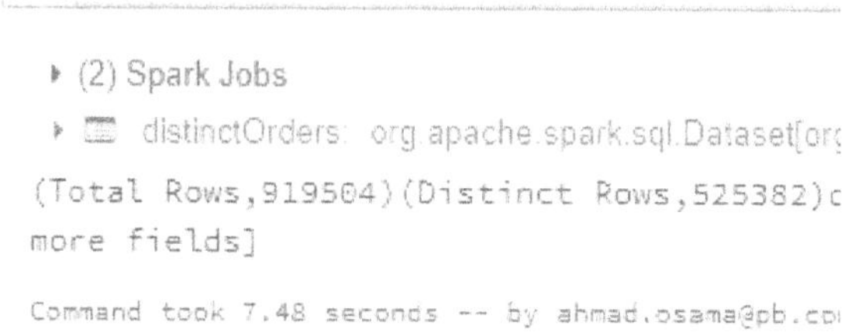

▶ (2) Spark Jobs

▶ ▦ distinctOrders org.apache.spark.sql.Dataset[org

(Total Rows,919504)(Distinct Rows,525382)c
more fields]

Command took 7.48 seconds -- by ahmad.osama@pb.com

Figure 9.39 – Removing duplicate records

The orders data is now ready for us to perform aggregation on it.

17. We'll calculate the total sales by customername and country. To do that, copy and paste the following code into a new cell:

```
// join distinctOrders and customer dataframe
val salesbycustomercountry = cleanorders
.join(customer,cleanorders("customerid")===customer
("id"),
"inner") // join cleanorders and customer dataframe on
customerid
.select(customer("name").alias("CustomerName"),
cleanorders("country"),(cleanorders("quantity")
*cleanorders("unitprice")).alias("TotalPrice")) //
select
relevant columns and add a new calculated column
TotalPrice = quantity*unitprice
.groupBy("CustomerName","country").agg(sum("TotalPrice")
.cast("decimal(10,2)").alias("Amount")) // calculate
total sales by customer and country
.sort(desc("Amount")) // sort on Amount descending order
display(salesbycustomercountry)
```

The preceding code joins the `cleanorders` and `customer` DataFrames, selects a name from the `customer` DataFrame (`CustomerName`), a country from the `cleanorders` DataFrame, and adds a new calculated column called `TotalPrice = quantity*unitprice`. Now, we must group the result set by the `CustomerName` and `country` columns and calculate the sum for each group.

You should get the following output:

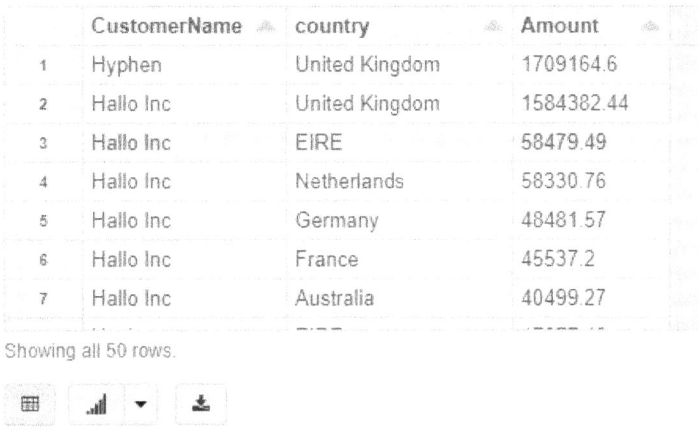

	CustomerName	country	Amount
1	Hyphen	United Kingdom	1709164.6
2	Hallo Inc	United Kingdom	1584382.44
3	Hallo Inc	EIRE	58479.49
4	Hallo Inc	Netherlands	58330.76
5	Hallo Inc	Germany	48481.57
6	Hallo Inc	France	45537.2
7	Hallo Inc	Australia	40499.27

Showing all 50 rows.

Figure 9.40 – Calculating sales by customer and country

18. Now, let's calculate the running total by country and invoice month. To do that, copy and paste the following code into a new cell:

```
import org.apache.spark.sql.functions._
import org.apache.spark.sql.types._
import org.apache.spark.sql.expressions.Window

val transform1 = \
cleanorders.select(cleanorders("country"),
// convert invoicedate to timestamp
unix_timestamp(cleanorders("invoicedate"),
            "yyyy-MM-dd HH:mm:ss")\
.cast(TimestampType).as("newinvoicedate"),
// calculate totalprice and convert to decimal(10,2)
(cleanorders("quantity")*cleanorders("unitprice"))\
```

```
                      .alias("TotalPrice").cast("decimal(10,2)"))
  .withColumn("invoicemonth",month($"newinvoicedate"))
                      // get month part from invoicedate
// partition by country and order by country and
invoicemonth

val defWindow =\
Window.partitionBy($"country").orderBy($"country",
$"invoicemonth")\
 .rowsBetween(Window.unboundedPreceding,0)
// apply the partition definition
val runningTotal = transform1.select(
                      $"country",$"invoicemonth",
                      row_number.over(defWindow).\
                              as("rownumber"),
                      sum($"TotalPrice").over(defWindow).\
                              as("runningtotal")
)

display(runningTotal)
```

The preceding code creates a DataFrame called `transform1` with a country called `newinvoidedate` and a `totalprice` column. We need to get the invoice month by applying the month function to `invoicedate`. However, the `invoicedate` column is of the `string` datatype. Therefore, we must convert `invoicedate` into `newinvoicedate`, which is of the `timestamp` datatype. We then get the invoice month.

19. To calculate the running total, we must use the `window` function to partition the DataFrame by country and order each partition by country and invoice month. To get the running total, we can use `rowsBetween (Window.UnboundedPrceding, 0)`, which further defines the window so that it includes all the rows from the start of the partition to the current row. We can then apply the `window` function to the `sum` function to get the running total. You should get the following output:

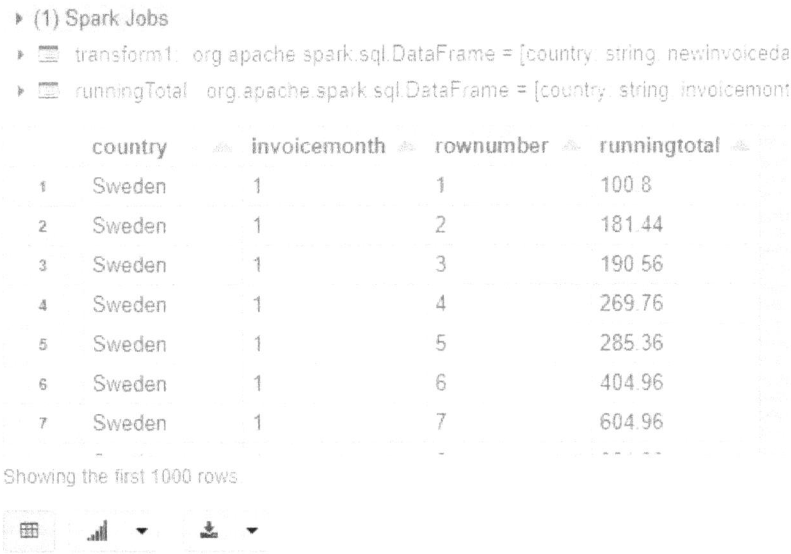

▸ (1) Spark Jobs

▸ 🗔 transform1: org.apache.spark.sql.DataFrame = [country: string, newinvoiceda

▸ 🗔 runningTotal: org.apache.spark.sql.DataFrame = [country: string, invoicemont

	country	invoicemonth	rownumber	runningtotal
1	Sweden	1	1	100.8
2	Sweden	1	2	181.44
3	Sweden	1	3	190.56
4	Sweden	1	4	269.76
5	Sweden	1	5	285.36
6	Sweden	1	6	404.96
7	Sweden	1	7	604.96

Showing the first 1000 rows

Figure 9.41 – Calculating the running total

20. Now, let's insert the result of the previous step, which is total sales by customer name and country, into the Azure Synapse SQL Pool. To do that, copy and paste the following code into a new cell:

```
import com.microsoft.azure.sqldb.spark.bulkcopy.
BulkCopyMetadata
import com.microsoft.azure.sqldb.spark.config.Config
import com.microsoft.azure.sqldb.spark.connect._
import com.microsoft.azure.sqldb.spark.query._
import org.apache.spark.sql.SaveMode

val truncateQuery = "TRUNCATE TABLE dbo.SalesStaging"
val queryConfig = Config(Map(
```

```
    "url"                  -> "azadesqlserver.database.
windows.net",
    "databaseName"        -> "azadesqldw",
    "user"                -> "sqladmin",
    "password"            -> "Sql@Server@1234",
    "queryCustom" -> truncateQuery
))
// truncate table SalesStaging

sqlContext.sqlDBQuery(queryConfig)

val insertConfig = Config(Map(
    "url"                  -> "azadesqlserver.database.
windows.net",
    "databaseName"        -> "azadesqldw",
    "user"                -> "sqladmin",
    "password"            -> "Sql@Server@1234",
    "dbTable"             -> "dbo.SalesStaging"
))

salesbycustomercountry.write.mode(SaveMode.Append).
sqlDB(insertConfig)
```

The preceding code truncates the `SalesStaging` table by running `TRUNCATE
TABLE` and then inserts the `salesbycustomercountry` DataFrame into the
`SalesStaging` table in Synapse SQL pool.

Note

Using `SaveMode.Overwrite` will result in an error. This is because the
`Overwrite` mode will drop and recreate the table with a `varchar(max)`
datatype for the `country` column. The `varchar(max)` method will give
us an error as it's not supported as a clustered column store index in an Synapse
SQL Pool.

Working with Delta Lake

Delta Lake is a layer between Spark and the underlying storage (Azure Blob Storage, Azure Data Lake Gen2) and provides **Atomicity, Consistency, Isolation, and Durability (ACID)** properties to Apache Spark. Delta Lake uses a transaction log to keep track of transactions and make transactions ACID compliant.

In this recipe, we'll learn how to perform insert, delete, update, and merge operations with Delta Lake, and then learn how Delta Lake uses transaction logs to implement ACID properties.

Getting ready

To get started, follow these steps:

1. Log into `https://portal.azure.com` using your Azure credentials.

2. You will need an existing Azure Databricks workspace and at least one Databricks cluster. You can create these by following the *Configuring an Azure Databricks environment* recipe.

How to do it...

Let's start by creating a new SQL notebook and ingesting data into Delta Lake:

> **Note**
>
> If you don't have time to follow these steps, import the `~/Chapter09/WorkingWithDeltaLake.dbc` notebook into your Databricks workspace.

1. In Azure portal, open the Azure Databricks service and click on **Launch Workspace** to open the Databricks plane.

2. In the Databricks plane, select **Workspace** from the left menu and create a new notebook with a default SQL language. Name the notebook `Working with Delta Lake`.

   In the previous recipe, we mounted the data lake container onto Databricks. If you are following on from the previous recipe, then you don't have to mount it again; if not, follow *Step 4* of the *Transforming data with Scala* recipe to mount the data lake container.

3. Copy and execute the following code in the notebook to fetch the orders from the
    `orders` folders into the `orders` DataFrame, and then create a global view to
    access the DataFrame data in the SQL scope:

```scala
%scala
// %scala allows us to run scala commands in the SQL
notebook.
// Similary we can switch to any language within a
notebook by specifying % and the language.


// load data into orders dataframe from Azure data lake.
// read data from orders container
import org.apache.spark.sql.types._

// specify the orders file schema.
// The orders text file don't have headers. If the file
has headers,
// we can use
// inferSchema to get column datatypes
val orderSchema = StructType(Array(
  StructField("invoiceno", StringType, true),
  StructField("stockcode", StringType, true),
  StructField("description", StringType, true),
  StructField("quantity", IntegerType, true),
  StructField("invoicedate", StringType, true),
  StructField("unitprice", DoubleType, true),
  StructField("customerid", IntegerType, true),
  StructField("Country", StringType, true),
  StructField("orderid", IntegerType, true))
)
// read all the text files in the orders container.
val orders = spark.read
.format("com.databricks.spark.csv")
.option("header","true")
.option("delimiter","|")
.schema(orderSchema)
.load("/mnt/datalake/orders/")
```

```
// create view to access the dataframe in sql context
orders.createOrReplaceTempView("orders");
```

> **Note**
>
> We can use magic commands to switch between languages; however, to access DataFrames between languages, we need to create views.

4. Once the `orders` view has been created, we can query it using a regular SQL command, as shown here:

```
-- query the orders view
select * from orders;
```

You should get the following output:

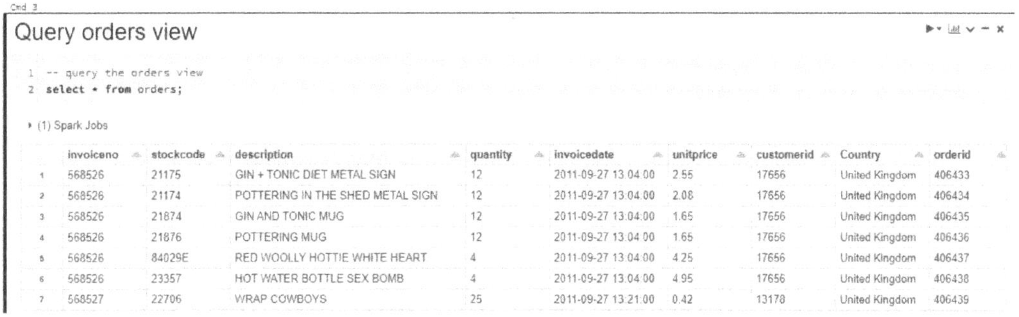
Cmd 3

Query orders view

```
1 -- query the orders view
2 select * from orders;
```

(1) Spark Jobs

	invoiceno	stockcode	description	quantity	invoicedate	unitprice	customerid	Country	orderid
1	568526	21175	GIN + TONIC DIET METAL SIGN	12	2011-09-27 13:04:00	2.55	17656	United Kingdom	406433
2	568526	21174	POTTERING IN THE SHED METAL SIGN	12	2011-09-27 13:04:00	2.08	17656	United Kingdom	406434
3	568526	21874	GIN AND TONIC MUG	12	2011-09-27 13:04:00	1.65	17656	United Kingdom	406435
4	568526	21876	POTTERING MUG	12	2011-09-27 13:04:00	1.65	17656	United Kingdom	406436
5	568526	84029E	RED WOOLLY HOTTIE WHITE HEART	4	2011-09-27 13:04:00	4.25	17656	United Kingdom	406437
6	568526	23357	HOT WATER BOTTLE SEX BOMB	4	2011-09-27 13:04:00	4.95	17656	United Kingdom	406438
7	568527	22706	WRAP COWBOYS	25	2011-09-27 13:21:00	0.42	13178	United Kingdom	406439

Figure 9.42 – Querying the orders view

5. Now, we'll create a database and a delta table from the `orders` view. To do this, copy and execute the following code in a new notebook cell:

```
-- Drop and Create a database in spark
Drop database if exists salesdwh cascade;
create database salesdwh;
-- Import data into delta lake from the orders dataframe
-- the following command creates a delta lake table
saleswh.orders_staging at the specified location.
-- the delta lake table will always have parquet format.
-- the staging table is dropped and created
CREATE TABLE salesdwh.orders_staging
USING delta
```

```
location '/mnt/ecommerce/salesdwh/orders_staging' -- this
location/folder in the ecommerce data lake container
should be empty
AS SELECT *
FROM orders;


-- query the saleswh.orders_staging table.
select * from salesdwh.orders_staging;
```

In the preceding command, we created a database called salesdwh to host all
the tables relevant to processing. If you don't create a database, the tables will be
created in the *default* database. We then used the CREATE TABLE AS SELECT
* command to create a delta table with data from the orders view. The USING
delta parameter specifies that the table is in delta format. We can specify the data's
location using the location command. The CREATE TABLE command creates
the /salesdwh/orders_staging path and copies the orders_staging data
to it in delta format.

> **Note**
>
> If the location isn't empty, the CREATE TABLE command may fail with an
> The associated location is not empty error. You'll have to
> remove the salesdwh folder and run the command again.

You should get the following output:

▶ (4) Spark Jobs

	invoiceno	stockcode	description	quantity	invoicedate	unitprice	customerid	Country	orderid
1	568526	21175	GIN + TONIC DIET METAL SIGN	12	2011-09-27 13:04:00	2.55	17656	United Kingdom	406433
2	568526	21174	POTTERING IN THE SHED METAL SIGN	12	2011-09-27 13:04:00	2.08	17656	United Kingdom	406434
3	568526	21874	GIN AND TONIC MUG	12	2011-09-27 13:04:00	1.65	17656	United Kingdom	406435
4	568526	21876	POTTERING MUG	12	2011-09-27 13:04:00	1.65	17656	United Kingdom	406436
5	568526	84029E	RED WOOLLY HOTTIE WHITE HEART.	4	2011-09-27 13:04:00	4.25	17656	United Kingdom	406437
6	568526	23357	HOT WATER BOTTLE SEX BOMB	4	2011-09-27 13:04:00	4.95	17656	United Kingdom	406438
7	568527	22706	WRAP COWBOYS	25	2011-09-27 13:21:00	0.42	13178	United Kingdom	406439

Figure 9.43 – Querying the delta table

With that, the data has been successfully loaded into the delta table from the
orders view.

6. To view the objects that were created as part of table creation, switch to Azure portal, open **packtdatalake** (Azure Data Lake Storage Gen2 account), and navigate to the ecommerce/salesdwh/orders_staging folder:

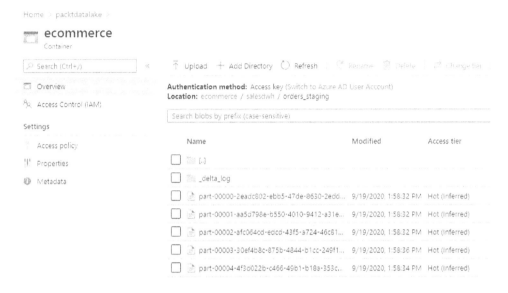

Figure 9.44 – Viewing delta lake objects

7. There are multiple parquet data files and a _delta_log directory for hosting the transaction log file. Open the _delta_log folder:

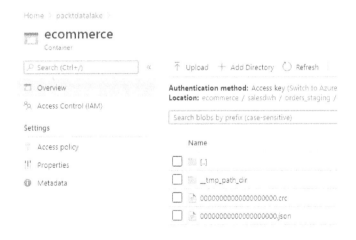

Figure 9.45 – Viewing transaction log files

The `_delta_log` folder contains a `.crc` file and a `.json` file. The `.json` file is a transaction log file. As the table will be updated, there'll be more `.json` files.

8. Open the `.json` file to view its content:

```json
{
  "commitInfo":
  {
    "timestamp":1600505491930,
    "userId":"301442434864740",
    "userName":"ahmad.osama@pb.com",
    "operation":"CREATE TABLE AS SELECT",
    "operationParameters":
    {
      "isManaged":"false",
      "description":null,
      "partitionBy":"[]",
      "properties":"{}"
    },
    "notebook":
    {
      "notebookId":"4271318757879724"
    },
    "clusterId":"0828-184032-doc671",
    "isolationLevel":"WriteSerializable",
    "isBlindAppend":true,
    "operationMetrics":
    {
      "numFiles":"5",
      "numOutputBytes":"10774220",
      "numOutputRows":"1616281"
    }
  }
}
```

The commitInfo section records the commit time, user, notebook, cluster, and other metadata information about the transaction. Also, observe that the data is automatically partitioned into multiple files. The subsequent entries list the file path and the meta data information about each file, such as the partition's start, number of rows, and more.

> **Note**
>
> Not all of the transaction log has been shown, for the sake of brevity.

9. Close the file. We'll look at the transaction log entries once we've modified the table, later in this recipe.

10. We can also use the COPY INTO command to directly insert the data into a delta table from the files. The COPY INTO command was in preview at the time of writing this book. Before we run the COPY INTO command, copy and execute the following command to drop the /mnt/ecommerce/salesdwh directory into a new cell:

```scala
%scala
dbutils.fs.rm("/mnt/ecommerce/salesdwh", true)
```

11. Copy and execute the following command to drop and create the salesdwh database and create the required staging and main table:

```
drop database if exists salesdwh cascade;
create database salesdwh;

-- create table orders_staging
CREATE TABLE IF NOT EXISTS salesdwh.orders_staging(
InvoiceNo string,
StockCode string,
Description string,
Quantity int,
InvoiceDate string,
UnitPrice string,
CustomerID string,
Country string,
orderid int
)
```

```
using delta
location '/mnt/ecommerce/salesdwh/orders_staging';

-- create table orders_staging1
CREATE TABLE salesdwh.orders_staging1(
_c0 string,
_c1 string,
_c2 string,
_c3 string,
_c4 string,
_c5 string,
_c6 string,
_c7 string,
_c8 string
)
using delta
location '/mnt/ecommerce/salesdwh/orders_staging1';

-- create table orders
CREATE TABLE IF NOT EXISTS salesdwh.orders(
InvoiceNo string,
StockCode string,
Description string,
Quantity int,
InvoiceDate string,
UnitPrice string,
CustomerID string,
Country string,
orderid int
)
using delta
location '/mnt/ecommerce/salesdwh/orders';
```

The preceding command creates the `orders_staging1`, `orders_staging` and `orders` delta tables.

12. Copy and execute the following command in a new cell to load the data into the orders_staging table using the COPY INTO command:

```
-- ingest data using the copy statement
-- only loads data from new files and skips older files
by default.
COPY INTO salesdwh.orders_staging1
   FROM '/mnt/ecommerce/orders'
   FILEFORMAT = CSV
   PATTERN = '*.txt*'
   FORMAT_OPTIONS ('delimiter'='|','header'='false');
insert into salesdwh.orders_staging select * from
salesdwh.orders_staging1;
select * from salesdwh.orders_staging;
```

The preceding command uses the COPY INTO command to load the data from the files to the orders_staging1 table. The FILEFORMAT command specifies the type of file, the PATTERN command specifies the type of files in the orders directory to read, and the FORMAT_OPTIONS command specifies the delimiter and the header.

13. Then, load the data from the orders_staging1 table into the orders_staging table using the INSERT INTO command.

If we directly load the data into the orders_staging table, we get a schema mismatch error as the column names in the orders_staging tables are different than those in the text files.

The COPY INTO command's default behavior is to skip the files that have already been read; however, we can use the FORCE command to read the data from the files that have already been read.

14. Now, we'll clean the orders_staging table and use the merge command to upsert the data into the orders table. Copy and execute the following command to clean and upsert the data into the orders table from the orders_staging table:

```
-- clean, transform and ingest the data into the orders
table.
-- 1. Update country "Unspecified" to "India"
-- 2. Update customerid to 1 where it's null
-- 3. Merge orders from orders_staging
```

```
   UPDATE salesdwh.orders_staging set Country='India' where
   Country='Unspecified';


-- 2. Update customerid to 1 where it's null
   delete from salesdwh.orders_staging where CustomerID is
   null;


-- 3. Merge orders from orders_staging
MERGE INTO salesdwh.orders as orders
USING salesdwh.orders_staging as staging
ON (orders.InvoiceNo = staging.InvoiceNo and orders.
Stockcode =
staging.StockCode and orders.CustomerID = staging.
CustomerID)
WHEN MATCHED
   THEN UPDATE SET orders.UnitPrice=staging.UnitPrice,
   orders.Quantity=staging.Quantity, orders.
Country=staging.Country
WHEN NOT MATCHED
   THEN INSERT (InvoiceNo, StockCode, Description,
Quantity, InvoiceDate,
            UnitPrice, CustomerID, Country, orderid)
   VALUES (InvoiceNo, StockCode, Description, Quantity,
InvoiceDate,
            UnitPrice, CustomerID, Country, orderid)
```

The preceding command updates the country to India, where the country is Unspecified, deletes the rows where customerid is null, and upserts the data into the orders table.

The upsert is done by using the Merge statement. The source for the Merge statement is orders_staging and its destination is the orders table. It updates the rows with matching InvoiceNo, Stockcode, and CustomerID and inserts the rows into the orders table. These are present in orders_staging but not in the orders table.

Let's review the `salesdwh` folder in the Azure Data Lake Storage account:

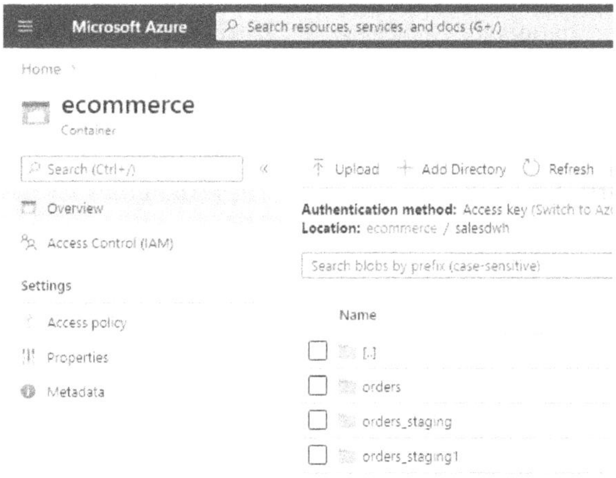

Figure 9.46 – Viewing the salesdwh folder in Azure Storage

There are three folders – one for each table we created earlier. Let's open the transaction log folder for the `orders_staging` table:

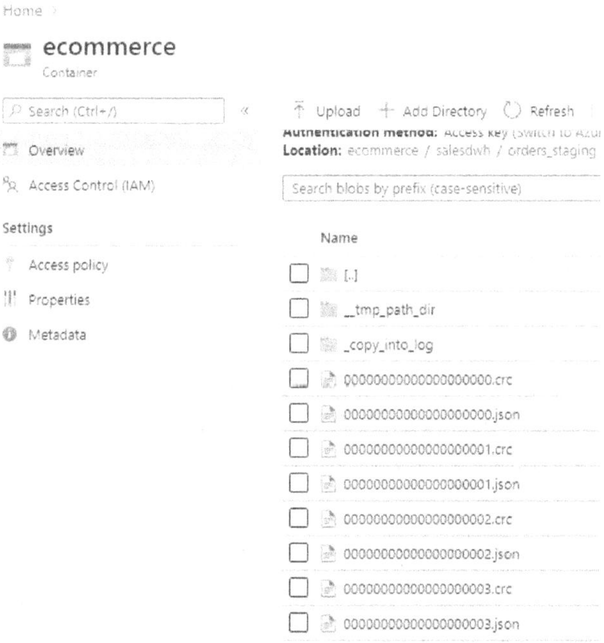

Figure 9.47 – Viewing the _delta_log folder

There are multiple transaction logs. Each transaction log corresponds to the operation's performance on the table. `00000000000000000000.json` contains the `commitInfo` details for the `CREATE TABLE` operation when we first created the table. `00000000000000000001.json` contains the `commitInfo` details for the write or the insert that we performed when inserting data from the orders_staging1 table into orders_staging. It also contains the path of the data files we created as part of the write information. These data files are as follows:

a) **File1**: `{"add":{"path":"part-00000-711886df-0476-47a9-a292-5ca201f743d2-c000.snappy.parquet"`

b) **File 2**: `{"add":{"path":"part-00001-e4213141-e125-42bc-a505-89608065130a-c000.snappy.parquet"`

c) **File 3**: `{"add":{"path":"part-00002-6598b911-0b80-4bdc-99a2-0400f9b40177-c000.snappy.parquet"`

> **Note**
> The filenames will be different in your case. The files are referred to as file 1, 2, and 3 for ease of understanding, and the files are not named like this in the transaction log file.

`00000000000000000002.json` contains the `commitInfo` details for the `Update` operation and updates the country to `India`, where the country is `Unspecified`. Other than the usual `add` entries (which specify data file paths), it also contains the `remove` entries with the data file path. It contains entries for removing or excluding the following data files:

a) **File 2**: `{"remove":{"path":"part-00001-e4213141-e125- -a505-89608065130a-c000.snappy.parquet","deletionTimestamp":1600516250452,"dataChange":true}}`

b) **File 1**: `{"remove":{"path":"part-00000-711886df-0476-47a9-a292-5ca201f743d2-c000.snappy.parquet","deletionTimestamp":1600516250452,"dataChange":true}}`

It contains entries so that you can add the following files:

a) **File 4**: `{"add":{"path":"part-00000-99218b0b-9ef5-4af9-9f08-0500026196fd-c000.snappy.parquet"`

b) **File 5**: `{ "add" : { "path" : "part-00001-de429c2a-6c5e-4d12-9f7a-f7366d2c9536-c000.snappy.parquet"`

File 1 and *File 2* from `00000000000000000001.json` are removed in the `00000000000000000002.json` log file and two new files are added as a result of the `Update` operation.

Any process querying the `orders_staging` table at the time of the `Update` operation will read from Files 1, 2, and 3 as the `00000000000000000002.json` log is not available yet.

Similarly, the `00000000000000000003.json` log is created after the `Delete` command is used. It also removes and adds the new files to the transaction log as required.

The `add` or `remove` entries in the transaction log file don't remove the physical data files. The entries tell Spark to skip the files with `remove` entries and to read from the files with `add` entries.

Processing structured streaming data with Azure Databricks

Streaming data refers to a continuous stream of data from one or more sources, such as IoT devices, application logs, and more. This streaming data can be either be processed record by record or in batches (sliding window) as required. A popular example of stream processing is finding fraudulent credit card transactions as and when they happen.

In this recipe, we'll use Azure Databricks to process customer orders as and when they happen, and then aggregate and save the orders in Azure Synapse SQL pool.

We'll simulate streaming data by reading the `orders.csv` file and sending the data row by row to an Azure Event hub. We'll then read the events from the Azure Event hub, before processing and storing the aggregated data in an Azure Synapse SQL pool.

Getting ready

To get started, follow these steps:

1. Log into `https://portal.azure.com` using your Azure credentials.

2. You will need an existing Azure Databricks workspace and at least one Databricks cluster. You can create these by following the *Configuring an Azure Databricks environment* recipe.

3. You will need an existing Azure Data lake Gen2 account mounted on Azure DBFS.

4. You will need an existing Azure storage account.

How to do it...

Let's started creating an Azure Event hub:

1. Execute the following PowerShell command to provision an Azure Event hub namespace and an Azure Event hub:

```
.\azure-data-engineering-cookbook\Chapter09\
CreateAzureEventHub.ps1 -resourcegroup packtade -location
centralus -eventhubnamespace packteventhub -eventhubname
ordersehub
```

The preceding command uses the `New-AzEventhubNamespace` cmdlet to create a new Azure Event hub namespace and `New-AzEventhub` to create an event hub in the event hub namespace. It then uses `Get-AzEventHubKey` to get the connection string and outputs the connection string to be used later.

You should get the following output:

```
PS C:\> .\azure-data-engineering-cookbook\chapter10\CreateAzureEventHub.ps1 -resourcegroup packtade -location centralus
-eventhubnamespace packteventhub -eventhubname ordersehub
WARNING: Breaking changes in the cmdlet 'New-AzEventHubNamespace' :
WARNING:   - "The output type 'Microsoft.Azure.Commands.EventHub.Models.PSNamespaceAttributes' is changing"
 - The following properties in the output type are being deprecated :
 'ResourceGroup'
 - The following properties are being added to the output type :
 'ResourceGroupName' 'Tags'

WARNING: NOTE : Go to https://aka.ms/azps-changewarnings for steps to suppress this breaking change warning, and other i
nformation on breaking changes in Azure PowerShell.
VERBOSE: Performing the operation "Create a new EvetntHub-Namespace:packteventhub under Resource Group:packtade" on targ
et "packteventhub".
Endpoint=sb://packteventhub.servicebus.windows.net/;SharedAccessKeyName=RootManageSharedAccessKey;SharedAccessKey=DxbTZL
LYmTeNxUk1JYhtFffh8M2skbz2/qBZHRvqjlw=;EntityPath=ordersehub
```

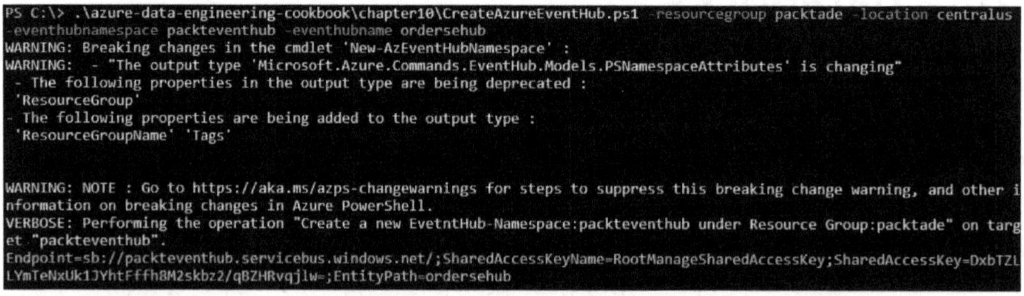

Figure 9.48 – Creating an Azure Event hub

Copy the connection string, as this will be used later.

2. Open `~/Chapter09/SendOrdersToEventHub/SendOrderstoEventHub.exe.config` in a notepad. Under **Configuration | appSettings**, replace the `eventhubname` and `connectionstring` values with your event hub and connection string, respectively. Replace the file path with the path of the `orders.csv` file. Save and close the configuration file. Then, double-click `~/Chapter09/SendOrdersToEventHub/SendOrderstoEventHub.exe` to start sending the orders from the `orders.csv` file to the Event hub. You should get the following output:

C:\azure-data-engineering-cookbook\Chapter10\SendOrdersToEventHub\SendOrderstoEventHub.exe
{"invoiceno":"536370","stockcode":"10002","description":"INFLATABLE POLITICAL GLOBE ","quantity":48,"invoicedate":"2010-12-01 08:45:00","unitprice":0.85,"productid":4,"country":"France","orderid":1467,"eventdatetime":"09-28-2020 10:18:30"}
{"invoiceno":"536382","stockcode":"10002","description":"INFLATABLE POLITICAL GLOBE ","quantity":12,"invoicedate":"2010-12-01 09:45:00","unitprice":0.85,"productid":4,"country":"India","orderid":1578,"eventdatetime":"09-28-2020 10:18:36"}
{"invoiceno":"536756","stockcode":"10002","description":"INFLATABLE POLITICAL GLOBE ","quantity":1,"invoicedate":"2010-12-02 14:23:00","unitprice":0.85,"productid":1,"country":"Bahrain","orderid":5708,"eventdatetime":"09-28-2020 10:18:37"}
{"invoiceno":"536863","stockcode":"10002","description":"INFLATABLE POLITICAL GLOBE ","quantity":1,"invoicedate":"2010-12-03 11:19:00","unitprice":0.85,"productid":7,"country":"Bahrain","orderid":6902,"eventdatetime":"09-28-2020 10:18:38"}
{"invoiceno":"536865","stockcode":"10002","description":"INFLATABLE POLITICAL GLOBE ","quantity":5,"invoicedate":"2010-12-03 11:28:00","unitprice":1.66,"productid":1,"country":"Bahrain","orderid":6982,"eventdatetime":"09-28-2020 10:18:39"}
{"invoiceno":"536876","stockcode":"10002","description":"INFLATABLE POLITICAL GLOBE ","quantity":2,"invoicedate":"2010-12-03 11:36:00","unitprice":1.66,"productid":1,"country":"Bahrain","orderid":7602,"eventdatetime":"09-28-2020 10:18:39"}

Figure 9.49 – Sending events to the Event hub

3. Next, we'll install the relevant library in Azure Databricks so that we can read events from the Event hub. To do that, open the Azure Databricks workspace in the Azure portal. From the right menu, select **Workspace**. Right-click anywhere on the **Workspace** tab, select **Create**, and then select **Library**.
4. On the **Create Library** window page, copy the `com.microsoft.azure:azure-eventhubs-spark_2.11:2.3.17` Maven coordinate into the **Coordinates** text box and click **Create**:

Microsoft Azure
Azure Databricks
Home
Workspace
Recents
Data
Create Library
Library Source
Upload
DBFS
PyPI
Maven
CRAN
Coordinates
com.microsoft.azure:azure-eventhubs-spark_2.11:2.3.17
Search Packages
Repository
Optional
Exclusions
Dependencies to exclude (log4j:log4j,junit:junit)
Create
Cancel

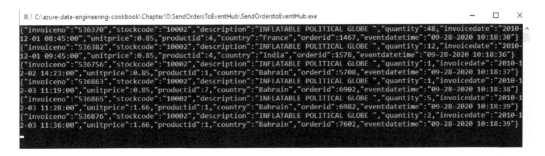

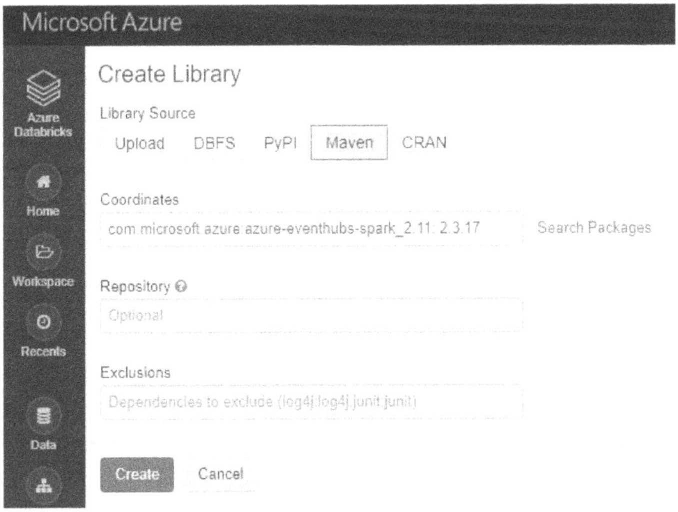

Figure 9.50 – Creating the library

The artifacts from the given coordinates will be listed:

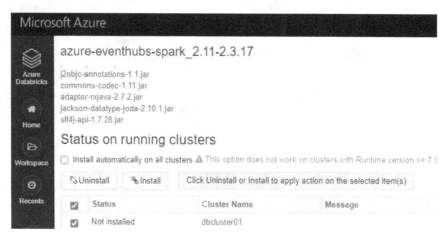

Figure 9.51 – Installing the library

5. Scroll down, select the Databricks cluster, and then click **Install** to install the library.

> **Note**
>
> The `com.microsoft.azure:azure-eventhubs-spark_2.11:2.3.17` library version works on Azure Databricks cluster 6.6 (including Apache Spark 2.4.5 and Scala 2.11). If you have a different version of the cluster, choose the appropriate library from here (**databricks** sections): `https://github.com/Azure/azure-event-hubs-spark`.

6. Now, let's create the notebook. To do that, create a new notebook with Python as the default language.

> **Note**
>
> If you don't have time to follow these steps, import the `~/Chapter09/ProcessStreamingData.dbc` notebook into your Databricks workspace.

7. Copy and paste the following code into a new cell in the notebook:

```
#Eventhub connection string
#Replace the connectionstring from step 1 - Recipe
("Process structured streaming data in Azure Databricks")
eventhubconnectionstring = "Endpoint=sb://
```

```
packteventhub.servicebus.windows.
net/;SharedAccessKeyName=fullaccess;SharedAccessKey=
ZRny+QSItU1fvIk5UG/
ugD7u1EHCpjHC7X+EC5WjxFY=;EntityPath=ordersehub"
# encrypt the connection string
eventhubconfig = {
   'eventhubs.connectionString' : sc._jvm.
org.apache.spark.eventhubs.EventHubsUtils.
encrypt(eventhubconnectionstring)
}
```

The preceding code configures the event hub connection string. Replace the value of the eventhubconnectionstring parameter with the value from *Step 1*.

8. Copy and paste the following command into a new cell to configure the Azure Synapse SQL Pool and Azure blob storage connection:

```
from pyspark.sql.functions import *
from pyspark.sql import *


# Azure SQL Server FQDN
jdbcHostname = "azadesqlserver.database.windows.net"
jdbcDatabase = "azadesqldw" # Azure SQL Database name
jdbcPort = 1433 # SQL Server default port
# get username and password for Azure SQL Database from
database secrets
username = dbutils.secrets.get(scope = "azuresql", key =
"username")
password = dbutils.secrets.get(scope = "azuresql", key =
"password")
# set the JDBC URL
jdbcUrl = \
"jdbc:sqlserver://
{0}:{1};database={2};user={3};password={4}".format(
    jdbcHostname, jdbcPort, jdbcDatabase, username,
password)


spark.conf.set(
  "fs.azure.account.key.packtstorage.blob.core.windows.
net",
```

```
"At2VRGhLB25rSeFlodCj/vr142AdzNSG8UcLVjomCb5V4iKPD/
v+4kuqI77tUVR3gtISFUB0gqitMoq0ezY32g==")
```

The preceding command configures the connection string for the Azure
Synapse SQL pool and Azure blob storage. Replace the parameter values as per
your environment.

9. Execute the following command in a new cell to read the events from the event hub:

```
# read events from Azure event hub
inputStreamdf = (
  spark
    .readStream
    .format("eventhubs")
    .options(**eventhubconfig)
    .load()
)
```

The preceding command gets the events into `inputStreamdf` in JSON format.

10. Execute the following command to convert the JSON into a defined
    structured schema:

```
from pyspark.sql.types import *
import  pyspark.sql.functions as F

# define the stream schema
ordersSchema = StructType([
  StructField("invoiceno", StringType(), True),
  StructField("stockcode", StringType(), True),
  StructField("description", StringType(), True),
  StructField("quantity", StringType(), True),
  StructField("invoicedate", StringType(), True),
  StructField("unitprice", StringType(), True),
  StructField("productid", StringType(), True),
  StructField("country", StringType(), True),
  StructField("orderid", StringType(), True),
  StructField("eventdatetime", StringType(), True),
  ])
```

```
#Apply the schema.
ordersdf = inputStreamdf.select(F.from_json(F.col("body")
.cast("string"), ordersSchema).alias("orders"))
# select the columns into a new dataframe
ordersStream = \
ordersdf.select("orders.invoiceno","orders.stockcode"
,"orders.description",
                "orders.quantity","orders.invoicedate",
                "orders.unitprice","orders.
productid","orders.country",
                "orders.orderid","orders.eventdatetime")
```

The preceding command defines the schema of the incoming streaming data. It then applies the schema to the JSON string and selects the relevant columns in the ordersStream DataFrame.

11. Execute the following command in a new cell to push the data to the memory sink:

```
# push data to memory sink for debugging
writeStreamtoMemory = (
  ordersStream \
    .writeStream \
    .format("memory") \
    .queryName("pushtomemorysink ")
    .start()
)
```

The preceding command sends the streaming data to the memory sink. We can insert the streaming data into a sink or a destination such as an Azure Synapse SQL Pool, Azure Data Lake Gen2, Cosmos DB, and more.

The memory sink and console sink are used to debug the stream. You should get an output similar to the following:

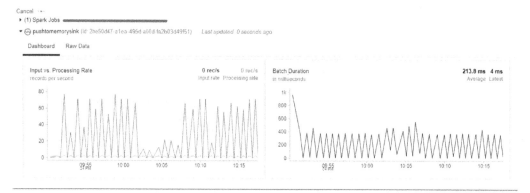

Figure 9.52 – Viewing the dashboard for a memory sink

The **Dashboard** tab provides a graph of input versus processing rate (records per second) and the batch's duration. This helps us identify whether we are receiving events and the rate at which we are receiving them.

12. Click **Cancel** to stop streaming to the memory sink.

13. Execute the following command to display the streaming DataFrame; that is, ordersStream:

```
# display the streaming dataframe
display(
        ordersStream,
        streamName="pushtomemorysink",
        processingTime = '10 seconds'
    )
```

The preceding command gets the data from the `pushtomemorysink` query in *Step 7*.

You should get an output similar to the following:

```
1  # display the streaming dataframe
2  display(ordersStream)
```

Cancel
▸ (1) Spark Jobs
▸ ⊙ display_query_2 (Id: c70e126e-064e-4bc4-8160-de517372ac5a)   Last updated: 5 seconds ago

	invoiceno	stockcode	description	quantity	invoicedate	unitprice	productid	country	orderid	eventdatetim
1	538517	18098C	PORCELAIN BUTTERFLY OIL BURNER	5	2010-12-12 16:05:00	2.95	7	Sweden	35007	09-28-2020
2	540464	18098C	PORCELAIN BUTTERFLY OIL BURNER	12	2011-01-07 13:17:00	2.55	7	Sweden	37467	09-28-2020
3	540462	18098C	PORCELAIN BUTTERFLY OIL BURNER	6	2011-01-07 12:45:00	2.95	7	Sweden	37420	09-28-2020
4	538827	18098C	PORCELAIN BUTTERFLY OIL BURNER	12	2010-12-14 12:59:00	2.55	7	Italy	40903	09-28-2020
5	538846	18098C	PORCELAIN BUTTERFLY OIL BURNER	2	2010-12-14 13:22:00	2.95	7	Italy	41110	09-28-2020
6	538671	18098C	PORCELAIN BUTTERFLY OIL BURNER	48	2010-12-13 16:38:00	2.55	7	Italy	39995	09-28-2020
7	538881	18098c	PORCELAIN BUTTERFLY OIL BURNER	3	2010-12-14 15:54:00	5.06	1	Italy	41870	09-28-2020

Figure 53 – Displaying the orderStream

> **Note**
>
> We can also run `display(ordersStream)` to view the data.

14. Execute the following command in a new cell to write the stream to a delta table:

```
# write stream to delta table
ordersOut = ordersStream \
.writeStream.format("delta") \
.option("checkpointLocation",
        "/mnt/ecommerce/checkpoint/orders_staging")\
.option("path", "/mnt/ecommerce/salesdwh/orders_
staging")\
.outputMode("append") \
.trigger(processingTime ='60 seconds')\
.start()
```

The preceding command writes the stream to a delta table at `/mnt/ecommerce/salesdwh/orders_staging`.

> **Note**
>
> In the *Transform data with Scala* recipe, we mounted an Azure Data Lake Gen2 filessystem called `ecommerce` on DBFS. If you don't have it mounted, please follow the steps in that recipe to mount the Azure Data Lake Gen2 filesystem on DBFS.

The delta format in `writeStream` specifies that it's a delta table. The `checkpointLocation` option specifies the checkpoint file's location. The checkpoint files contain the start and end offsets of each batch. In the case of failure, the checkpoint files are used to reprocess the data based on the batch offsets. The `trigger` option specifies the interval in which we can get the data. It's recommended to use `trigger` for development so as to reduce resource overhead. If the trigger isn't specified, the query gets the data as soon as it can. The `outputMode` option adds the new data to the table without modifying the existing data.

15. Hit *Shift + Enter* to run the query. Click **Cancel** to stop the query after a minute or two.

16. Execute the following query to get the data from the delta table:

```sql
%sql
select count(*) from delta.`/mnt/ecommerce/salesdwh/
orders_staging`
```

You should get an output similar to the following:

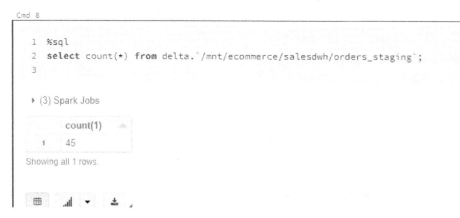

Figure 9.54 – Querying the orders_staging delta table

The count will differ in your case.

17. Now, we'll stream the data from the `orders_staging` delta table and then apply transformations to it. To do that, execute the following query in a new cell:

```
# stream the data from the delta table.
ordersdf = spark.readStream\
    .format("delta")\
    .load("/mnt/ecommerce/salesdwh/orders_staging")
```

```
from pyspark.sql.functions import *
from pyspark.sql.types import *


ordersTransformdf = ordersdf \
.filter((col("country")!="Unspecified") &
(col("quantity")>0)) \
.withColumn("totalamount", col("unitprice")*col
("quantity")) \
.withColumn("currenttimestamp",current_timestamp()) \
.withColumn("eventdatetime_ts",to_timestamp
("eventdatetime", "MM-dd-yyyy HH:mm:ss")) \
.select("invoiceno","stockcode","description","quantity",
"invoicedate",
        "unitprice","productid","country","orderid",
"eventdatetime_ts",
        "currenttimestamp","totalamount")
```

In the preceding command, we read the streaming data from the orders_
staging table into the ordersdf DataFrame. We then applied the following
transformations:

a) Filter out rows where country is Unspecified

b) Filter out rows where quantity is less than zero

c) Adds the totalamount column, which is a product of quantity and
unitprice

d) Add a new column called currenttimestamp with a default value of the
current date and time

e) Convert eventdatetime to timestamp datatype and add as new column,
eventdatetime_Ts

f) Select required columns

The transformed data is stored in the ordersTransformed DataFrame.

18. Now, let's aggregate the transformed data to calculate the sales by country. To do
this, execute the following query in a new cell:

```
import pyspark.sql.functions as fun
from pyspark.sql.functions import current_timestamp
```

```
# aggregate data
salesbycountrydf  = ordersTransformdf \
.withWatermark("eventdatetime_ts", "11 minutes")\
.groupBy(window("eventdatetime_ts", "10
minutes"),"country")\
.agg(fun.sum("totalamount").alias("Amount"),
     fun.min("eventdatetime_ts").alias("min_
eventdatetime_ts"),
     fun.max("eventdatetime_ts").alias("max_
eventdatetime_ts"))\
.select("country","Amount","min_eventdatetime_ts","max_
eventdatetime_ts")
```

In the preceding command, we calculated the total sales by country and stored the data in the `salesbycountrydf` DataFrame.

The `withWatermark` option specifies the time duration to preserve the state of the aggregation. In a streaming environment, events may arrive out of order; for example, an order generated at 12:00 p.m. may arrive after the orders generated at 12:05 p.m. To cater to such scenarios, the state of the aggregate is to be maintained so that the values can be updated as per the late events. However, having an indefinite state will result in resource exhaustion and performance degradation. The `withWatermark` option of 10 minutes will keep the data for the last 10 minutes, and any events arriving 10 minutes late will be added to the aggregate value; however, events later than 10 minutes will be dropped.

The watermark value may vary from system to system.

19. Hit *Shift + Enter* to run the query.

20. Execute the following query in a new cell to insert the aggregated data from *Step 11* into a new delta table:

```
# write aggregate to delta table
aggregatesOut = salesbycountrydf \
.writeStream.format("delta") \
.option("checkpointLocation",
        "/mnt/ecommerce/checkpoint/salesbycountry/") \
.option("path", "/mnt/ecommerce/salesdwh/
salesbycountry")\
.outputMode("append") \
.start()
```

In the preceding command, we wrote the aggregated data into the salesbycountry delta table. The checkpoint location for each write operation should be different.

21. Execute the following query in a new cell to write the aggregated delta table, salesbycountry, into an Azure Synapse SQL pool:

```
from pyspark.sql.functions import *
from pyspark.sql import *

# stream the data from salesbycountry delta table.
readsalesbycountry = spark.readStream \
.format("delta")\
.load("/mnt/ecommerce/salesdwh/salesbycountry")

# sink the data into Azure synapse SQL pool
readsalesbycountry.writeStream\
.format("com.databricks.spark.sqldw")\
.option("url", jdbcUrl)\
.option("forwardSparkAzureStorageCredentials", "true")\
.option("dbTable", "salesbycountry")\
.option("tempDir",
"wasbs://tempdir@packtstorage.blob.core.windows.net/
sqldw/salesbycountry")\
.trigger(processingTime="10 seconds")\
.option("checkpointLocation",
        "/mnt/ecommerce/checkpoint/sqldw/
salesbycountry")\
.outputMode("append")\
.start()
```

In the preceding command, first, we read the streaming data from the salesbycountry delta table into the readsalesbycountry DataFrame. We then wrote the streaming data into the salesbycountry Azure Synapse SQL Pool table.

22. Execute the following query to create the orders delta table:

```
%sql
drop table if exists orders;
```

```
CREATE TABLE if not exists orders(
    invoiceno string,
    stockcode string,
    description string,
    quantity int ,
    invoicedate timestamp,
    unitprice decimal(10,2),
    productid int,
    country string,
    orderId int,
    eventdatetime timestamp
    )
USING delta
location '/mnt/ecommerce/salesdwh/orders'
```

The preceding command creates a new delta table orders at /mnt/ecommerce/salesdwh/orders.

23. Execute the following command in a new cell to upsert the data into the orders table from the orders_staging table:

```scala
%scala
import org.apache.spark.sql._
import org.apache.spark.sql.functions._

// function to upsert the data for a batch into the delta table
def upsertToDelta(microBatchOutputDF: DataFrame, batchId: Long) {
  microBatchOutputDF.createOrReplaceTempView("updates")

  microBatchOutputDF.sparkSession.sql(s"""
    MERGE INTO orders t
    USING updates s
    ON s.orderid = t.orderid
    WHEN MATCHED THEN UPDATE SET *
    WHEN NOT MATCHED THEN INSERT *
    """)}
```

```
// read orders_staging delta table and call upsertToDelta
function for each batch
spark.readStream
  .format("delta")
  .load("/mnt/ecommerce/salesdwh/orders_staging")
  .writeStream
  .format("delta")
  .foreachBatch(upsertToDelta _)
  .outputMode("update")
  .start()
```

First, the preceding query defines a function, upserToDelta, in order to upsert the microBatchOutputDF DataFrame into the orders table. The function uses the Merge command to upsert the data based on the orderid column. It then reads the streaming data from the orders_staging table and calls the upsertToDelta function to merge the data into the orders table.

`Packt.com`

Subscribe to our online digital library for full access to over 7,000 books and videos, as well as industry leading tools to help you plan your personal development and advance your career. For more information, please visit our website.

Why subscribe?

- Spend less time learning and more time coding with practical eBooks and Videos from over 4,000 industry professionals

- Improve your learning with Skill Plans built especially for you

- Get a free eBook or video every month

- Fully searchable for easy access to vital information

- Copy and paste, print, and bookmark content

Did you know that Packt offers eBook versions of every book published, with PDF and ePub files available? You can upgrade to the eBook version at `packt.com` and as a print book customer, you are entitled to a discount on the eBook copy. Get in touch with us at `customercare@packtpub.com` for more details.

At `www.packt.com`, you can also read a collection of free technical articles, sign up for a range of free newsletters, and receive exclusive discounts and offers on Packt books and eBooks.

Other Books You May Enjoy

If you enjoyed this book, you may be interested in these other books by Packt:

Data Engineering with Python

Paul Crickard

ISBN: 978-1-83921-418-9

- Understand how data engineering supports data science workflows

- Discover how to extract data from files and databases and then clean, transform, and enrich it

- Configure processors for handling different file formats as well as both relational and NoSQL databases

- Find out how to implement a data pipeline and dashboard to visualize results

- Use staging and validation to check data before landing in the warehouse

- Build real-time pipelines with staging areas that perform validation and handle failures

- Get to grips with deploying pipelines in the production environment

Python Data Cleaning Cookbook

Michael Walker

ISBN: 978-1-80056-566-1

- Find out how to read and analyze data from a variety of sources
- Produce summaries of the attributes of data frames, columns, and rows
- Filter data and select columns of interest that satisfy given criteria
- Address messy data issues, including working with dates and missing values
- Improve your productivity in Python pandas by using method chaining
- Use visualizations to gain additional insights and identify potential data issues
- Enhance your ability to learn what is going on in your data
- Build user-defined functions and classes to automate data cleaning

Packt is searching for authors like you

If you're interested in becoming an author for Packt, please visit `authors.packtpub.com` and apply today. We have worked with thousands of developers and tech professionals, just like you, to help them share their insight with the global tech community. You can make a general application, apply for a specific hot topic that we are recruiting an author for, or submit your own idea.

Leave a review - let other readers know what you think

Please share your thoughts on this book with others by leaving a review on the site that you bought it from. If you purchased the book from Amazon, please leave us an honest review on this book's Amazon page. This is vital so that other potential readers can see and use your unbiased opinion to make purchasing decisions, we can understand what our customers think about our products, and our authors can see your feedback on the title that they have worked with Packt to create. It will only take a few minutes of your time, but is valuable to other potential customers, our authors, and Packt. Thank you!

Index